ATE DUE

VOLUME EIGHTY TWO

ADVANCES IN
GENETICS

ADVANCES IN GENETICS, VOLUME 82

Serial Editors

Theodore Friedmann
University of California at San Diego,
School of Medicine, CA, USA

Jay C. Dunlap
The Geisel School of Medicine at Dartmouth,
Hanover, NH, USA

Stephen F. Goodwin
University of Oxford, Oxford, UK

> VOLUME EIGHTY TWO

ADVANCES IN
GENETICS

Edited by

Theodore Friedmann
Department of Pediatrics,
University of California at San Diego,
School of Medicine, CA, USA

Jay C. Dunlap
Department of Genetics,
The Geisel School of Medicine at Dartmouth,
Hanover, NH, USA

Stephen F. Goodwin
Department of Physiology,
Anatomy and Genetics,
University of Oxford,
Oxford, UK

AMSTERDAM • BOSTON • HEIDELBERG • LONDON
NEW YORK • OXFORD • PARIS • SAN DIEGO
SAN FRANCISCO • SINGAPORE • SYDNEY • TOKYO
Academic Press is an imprint of Elsevier

ELSEVIER

Academic Press is an imprint of Elsevier
225 Wyman Street, Waltham, MA 02451, USA
525 B Street, Suite 1800, San Diego, CA 92101-4495, USA
Radarweg 29, PO Box 211, 1000 AE Amsterdam, The Netherlands
The Boulevard, Langford Lane, Kidlington, Oxford, OX5 1GB, UK
32 Jamestown Road, London, NW1 7BY, UK

First edition 2013

ISBN: 978-0-12-407676-1
ISSN: 0065-2660

For information on all Academic Press publications
visit our website at store.elsevier.com

Printed and bound in USA
13 14 15   10 9 8 7 6 5 4 3 2 1

Working together
to grow libraries in
developing countries

www.elsevier.com • www.bookaid.org

CONTENTS

CONTRIBUTORS

Jesper B. Bramsen
Interdisciplinary Nanoscience Center (iNANO), Department of Molecular Biology and Genetics, Aarhus University, Aarhus, Denmark

Aaron A. Goodarzi
Department of Biochemistry & Molecular Biology, and Department of Oncology, Southern Alberta Cancer Research Institute, University of Calgary, Calgary, Alberta, Canada

Bo R. Hansen
Santaris Pharma A/S, Hørsholm, Denmark

Torben Højland
NanoCAN, Department of Physics, Chemistry and Pharmacy, University of Southern Denmark, Odense, Denmark

Penelope A. Jeggo
Genome Damage and Stability Centre, University of Sussex, Brighton, Sussex, United Kingdom

Jørgen Kjems
Interdisciplinary Nanoscience Center (iNANO), Department of Molecular Biology and Genetics, Aarhus University, Aarhus, Denmark

Troels Koch
Santaris Pharma A/S, Hørsholm, Denmark

Karin E. Lundin
Clinical Research Center, Department of Laboratory Medicine, Karolinska Institutet, Novum, Huddinge, Stockholm, Sweden

Robert Persson
Santaris Pharma A/S, Hørsholm, Denmark

C.I. Edvard Smith
Clinical Research Center, Department of Laboratory Medicine, Karolinska Institutet, Novum, Huddinge, Stockholm, Sweden

Jesper Wengel
NanoCAN, Department of Physics, Chemistry and Pharmacy, University of Southern Denmark, Odense, Denmark

CHAPTER ONE

The Repair and Signaling Responses to DNA Double-Strand Breaks

Aaron A. Goodarzi[*,†], Penelope A. Jeggo[‡,1]

[*]Department of Biochemistry & Molecular Biology, Southern Alberta Cancer Research Institute, University of Calgary, Calgary, Alberta, Canada
[†]Department of Oncology, Southern Alberta Cancer Research Institute, University of Calgary, Calgary, Alberta, Canada
[‡]Genome Damage and Stability Centre, University of Sussex, Brighton, Sussex, United Kingdom
[1]Corresponding author: e-mail address: p.a.jeggo@sussex.ac.uk

Contents

Advances in Genetics, Volume 82
ISSN 0065-2660
http://dx.doi.org/10.1016/B978-0-12-407676-1.00001-9

Abstract

A DNA double-strand break (DSB) has long been recognized as a severe cellular lesion, potentially representing an initiating event for carcinogenesis or cell death. The evolution of DSB repair pathways as well as additional processes, such as cell cycle checkpoint arrest, to minimize the cellular impact of DSB formation was, therefore, not surprising. However, the depth and complexity of the DNA damage responses being revealed by current studies were unexpected. Perhaps the most surprising finding to emerge is the dramatic changes to chromatin architecture that arise in the DSB vicinity. In this review, we overview the cellular response to DSBs focusing on DNA repair pathways and the interface between them. We consider additional events which impact upon these DSB repair pathways, including regulated arrest of cell cycle progression and chromatin architecture alterations. Finally, we discuss the impact of defects in these processes to human disease.

1. INTRODUCTION

A first review entitled *DNA Breakage and Repair for Advances in Genetics* was written in 1998. At this time, the major focus was to describe and discuss the relatively newly identified genes encoding proteins that function in the major DNA double-strand break (DSB) repair pathway: DNA non-homologous end-joining (NHEJ). This was an exciting time in the field with the emerging notion that the major DSB repair pathway in mammalian cells was distinct to homologous recombination (HR), the process predominantly exploited by lower organisms, gaining credence as the NHEJ genes were identified. Additionally, an unanticipated role for NHEJ during V(D)J recombination, a critical process in development of the immune response which rearranges and rejoins the subexon components of the immunoglobulin and T cell receptor genes, had been newly revealed. Thus, understanding the NHEJ process was of interest not just to those workers in the DNA damage response (DDR) field but additionally to immunologists. Now in 2013, together with Aaron Goodarzi, we venture to revisit this area. Although distinct to earlier years, the field remains dynamic, rapidly advancing, and exciting. We now have substantial insight into the basic processes of NHEJ and HR at a structural and biochemical level and a sound appreciation of their cellular roles. Moreover, a recently described process of Alternative-NHEJ (Alt-NHEJ) has been identified. The significance of DSB signal transduction responses has gained credence and insight into the process and its impact on the DDR is slowly emerging. Critical current questions are: how these responses interface, how they are influenced by chromatin structure, and how chromatin is changed to optimally promote DSB repair and avoid genomic instability? Most provocatively, how does the malfunction of these processes lead to chromosomal translocations and rearrangements? The field is substantially broader and more complex than in 1998, making the task of providing an overview more challenging. Here, we aim to provide a description of our current understanding of the DSB repair processes, the signaling response, the interplay between them and other metabolic processes involving DNA, and the impact of, and changes to, the chromatin environment.

2. FORMATION OF DSBs

There is increasing recognition that the route by which a DSB arises strongly influences the pathway governing its repair. DSBs can arise in a developmentally programmed manner (such as V(D)J recombination, class switch recombination (CSR), or meiosis), following replication fork arrest or stalling, from endogenously arising DNA damage or from exogenous DNA-damaging agents such as ionizing radiation (IR). Each process produces DSBs of a distinctive nature (summarized in Figure 1.1). DSBs that arise following replication fork collapse are described frequently as one-ended since, due to the replication fork structure, they do not have a partner end (Figure 1.1). With no second end for rejoining, repair by NHEJ is problematic and there is a defined process to preclude Ku binding, the initiating step of NHEJ (see below), at replication associated DSBs which are repaired preferentially by HR (Adamo et al., 2010; Petermann & Helleday, 2010). However, in certain situations, recovery from replication stalling can involve new origin firing and NHEJ may then repair any ensuing two-ended DSB (Blow, Ge, & Jackson, 2011). Exogenous DNA-damaging agents such as IR can directly or indirectly (via replication) induce DSBs. IR-induced DSBs can be highly complex, particularly if they are induced by high-linear energy transfer (LET) radiation. DSBs can also arise following encounter of the transcription machinery with base damage or single-strand breaks (SSBs) or via a route involving topoisomerase II.

3. MECHANISMS OF DSB REJOINING

HR and NHEJ represent the two major DSB repair pathways; additionally, less well-understood processes have also been described.

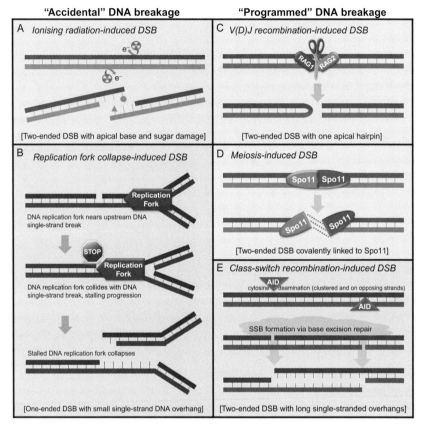

Figure 1.1 Variable routes of DSB formation: (A) Ionizing radiation can directly induce DSBs without replication or cellular processing. IR-induced DSBs have two termini (hence referred to as two-ended) and can have damaged bases or sugars at their termini and single-strand breaks in close proximity. (B) Replication forks can collapse when they encounter base damage or a single-strand break. Depending on whether the lesion is on the 5′ or 3′ strand, this can lead to single termini DSBs, that is, one ended. The ends can be a blunt end or, following resection, can have a 3′ overhang; the lesion on the other strand is a single-strand gap or region, depending on polarity. (C) During V(D)J recombination, the RAG1/RAG2 endonucleases cleave DNA at recombination signal sequences. Breakage occurs via a transesterification reaction generating one blunt and one hairpin-ended DSB. (D) During meiosis, the topoisomerase II-like protein, Spo11, and the MRN complex generate protein-bridged DSBs. (E) During class-switch recombination (CSR), AID (activation-induced cytosine deaminase) generates uracil residues by deamination of cytosines at CSR sites at which SSBs can arise following their processing by uracil glycosylase. Two closely localized SSBs degenerate into DSBs with long single-strand overhangs. (For color version of this figure, the reader is referred to the online version of this chapter.)

3.1. Homologous recombination

HR represents an exquisitely elegant process for DSB rejoining as it relies on undamaged templates to restore any lost sequence information. Given the diploid karyotype of mammalian cells, it might have been anticipated that HR could function throughout the cell cycle if the homologous chromosome can be used as a template. However, HR actually only functions in late S/G2 phase, the stages when a sister chromatid is available (Johnson & Jasin, 2000). This is likely because polymorphisms or homologue sequence differences, particularly in intronic regions, are too great to allow efficient heteroduplex formation. Importantly, although HR contributes to two-ended DSB repair in G2 phase cells, it functions mainly at the replication fork promoting fork restart and/or one-ended DSB repair (Adamo et al., 2010; Petermann & Helleday, 2010). Several outstanding reviews on HR have been written, and only a brief overview of more recently described aspects will be given here (Kass & Jasin, 2010; Mazon, Mimitou, & Symington, 2010; Moynahan & Jasin, 2010). The well-described steps of HR are (i) 3′ single-stranded (ss) DNA tail generation, a prerequisite for strand invasion and heteroduplex DNA formation; (ii) ssDNA coating with RPA; (iii) RPA displacement by RAD51 via a process involving BRCA2 to form a nucleoprotein filament; (iv) formation of heteroduplex DNA and Holliday junction (HJ) formation; (v) branch migration, and (vi) resolution, which, depending on the orientation can, but may or may not, lead to DNA-strand crossover (Figure 1.2).

The importance of regulating resection, which represents the major step determining the choice between NHEJ and HR, has been increasingly recognized (reviewed in Symington & Gautier, 2011). The prevailing evidence suggests that resection involves two distinct steps, an initiating event involving Sae2/CtIP and Mre11/MRE11, followed by an elongation step involving exonuclease I (ExoI) or a complex of Sgs1–Top3–RMI1 (STR)/DNA2/BLM. Important insight into the resection process has come from the study of meiosis in yeast, where a DSB is introduced by Spo11, a topoisomerase II-like enzyme, which generates a DSB with a 5′ end covalently bridged to a Spo11 tyrosine residue (Borde & de Massy, 2013). During meiosis, resection is initiated by an Mre11-mediated endonucleolytic incision on the 5′ strand, releasing a Spo11-bound 12–34 bp oligonucleotide (Garcia, Phelps, Gray, & Neale, 2011). It has been proposed that Mre11 exonuclease activity subsequently digests 3′–5′ toward the DNA end while Exo1 digests 5′–3′ away from the end (Garcia et al., 2011;

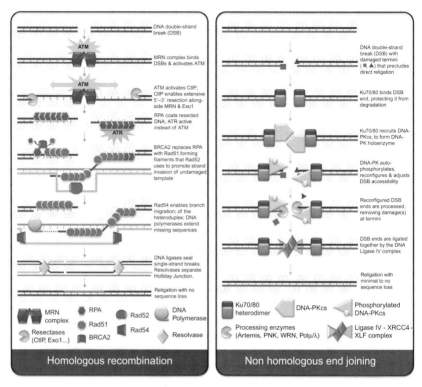

Figure 1.2 Major mechanisms of DSB repair: Left Panel, red: During HR, resection is initiated by MRN and CtIP and progressed by exonucleases such as ExoI and BLM/DNA2. The single-strand DNA generated is coated with RPA, which can activate ATR. Via BRCA2 and RAD52, RPA is displaced by RAD51 to generate DNA filaments which are capable of invading the undamaged strand to generate D-loops. The hybrid DNA generated is called heteroduplex DNA with junctions formed by DNA-strand crossing over referred to as Holliday junctions. The cross-over point can move by branch migration mediated by Rad54. Finally, resolvases separate Holliday junctions and generate products with or without crossovers, depending on the relative direction of resolution of the two Holliday junctions. Right panel, blue: During NHEJ the Ku heterodimer threads onto the DSB apex, encircling the DNA strand. DSB-bound Ku recruits DNA-PKcs, generating the DNA-dependent protein kinase (DNA-PK) holoenzyme, whose autophosphorylation permits end processing by a variety of factors prior to recruitment of a ligation complex involving DNA ligase IV, XRCC4, and XLF. This process occurs predominantly independently of ATM signaling. (For interpretation of the references to color in this figure legend, the reader is referred to the online version of this chapter.)

Symington & Gautier, 2011). However, it is likely that there are distinctions between the progression of resection during HR in meiosis versus mitosis due to the nature of the complexes bound at the DSB (Spo11 in meiosis and Ku in mitosis). The regulation of resection in mitotic cells, moreover, involves additional components, including BRCA1. This will be discussed further when considering the interplay between HR and NHEJ.

Our understanding of nucleoprotein filament formation and strand exchange during HR has been progressed by elegant studies including scanning force microscopy, single molecule experiments, and the visualization of DNA–protein interactions (Forget & Kowalczykowski, 2012; Hilario & Kowalczykowski, 2010; Holthausen et al., 2011). The assembly of RAD51 filaments is a dynamic process occurring by nucleation and subsequent growth of an ssDNA-bound protein filament. ATP hydrolysis drives the process, promoting subsequent rearrangement of RAD51 filaments to yield a joint molecule complex, promoting strand exchange and heteroduplex formation. BRCA2, which facilitates this process, directly interacts with RAD51 via BRCA2's BRCT repeats and a C-terminal RAD51-interaction domain (Jensen, Carreira, & Kowalczykowski, 2010). Although the precise mechanism remains unclear, one potential model is that BRCA2 interacts with RAD51–dsDNA complexes modulating RAD51–DNA binding to promote filament stability on ssDNA versus dsDNA (Holthausen et al., 2011). There is also evidence that the BRCT repeats fall into two classes, which differ in their precise function, but coordinate to promote efficient nascent RAD51 filament formation on ssDNA by enhancing both nucleation and subsequent filament growth (Carreira & Kowalczykowski, 2011). Interestingly, more recent studies have recognized that RAD51 nucleation takes places on RPA-coated ssDNA rather than naked DNA. While such analysis has not been carried out yet with human RAD51, nucleation of RecA on ssDNA-coated with single-strand binding protein, the bacterial equivalent of RPA, has been imaged (Bell, Plank, Dombrowski, & Kowalczykowski, 2012). In addition to BRCA2, RAD54, a dsDNA-dependent ATPase and a member of the Snf2 family also impact upon RAD51 nucleoprotein filament formation, both stabilizing and disrupting such filaments (Mazin, Mazina, Bugreev, & Rossi, 2010). While the impact of RAD54 is multiple, its precise function in HR remains unclear.

The branch migration step of HR represents a process by which one DNA strand is progressively exchanged for another. This is a motor-driven process with several helicase-like proteins being involved; these include BLM, WRN (Werner's syndrome protein), RECQ1, RECQ5, and RAD54, although their interplay *in vivo* remains unclear (Larsen & Hickson, 2013). Substantial insight into the resolution step of HR has been gained in recent years, with three pathways promoting HJ resolution/dissolution described (Wechsler, Newman, & West, 2011); interplay between these pathways appears complex, with distinctions between organisms. Here, we focus on mammalian cells, where the major way to promote resolution of the so-called HJs, the point of crossing over of the DNA strands, is via symmetric nicking at the base of the junction. After many years of searching, GEN1, homologous to yeast Yen1, was identified as being able to resolve HRs by a RuvC-like symmetrical cleavage (Ip et al., 2008). Subsequently, MUS81–EME1, a member of the XPF family of heterodimeric nucleases, was also shown to resolve recombination intermediates (Wechsler et al., 2011). Initial studies suggested that MUS81–EME1 preferentially cleaves $3'$-flap and nicked HJs, and that intact four-way junctions are a poor substrate. Additionally, SLX4 functions as a scaffolding protein for the assembly of multiple endonucleases, which together can promote HJ resolution (Svendsen & Harper, 2010); whether SLX4 functions with the other identified resolvases remains to be examined. An alternative process of HJ dissolution (versus resolution) has also been described, involving BLM, topoisomerase IIIa, RMI1, and RMI2 (Wu & Hickson, 2003). A recent study provided evidence that resolution is subject to regulatory control and that different processes function at specific stages in meiotic and mitotic HR, providing at least a partial explanation for these overlapping pathways (Matos, Blanco, Maslen, Skehel, & West, 2011).

3.2. Nonhomologous end-joining

NHEJ has also been reviewed previously and our focus will rest on the more recent advances (Lieber, 2010; Neal & Meek, 2011). NHEJ functions throughout the cell cycle and represents the major mechanism for repairing two-ended DSBs such as those generated by IR. Indeed, mutants deficient in NHEJ proteins display dramatic radiosensitivity.

3.2.1 DNA-PK: The first complex formed during NHEJ

NHEJ is initiated by the avid ability of the Ku70–Ku80 (Ku) heterodimer, to bind double-stranded (ds) DNA ends. Structural analysis of Ku has revealed a basket-shaped molecule with a core that allows threading onto a dsDNA ends without sequence specificity, making contact with the DNA backbone but not bases (Walker, Corpina, & Goldberg, 2001). Additionally, the structure of Ku suggested a ratchet-type mechanism that allows it to translocate along the DNA molecule, a feature previously observed from biochemical studies. Importantly, Ku binding to dsDNA protects DNA ends from nucleolytic degradation and recruits the DNA-PK catalytic subunit (DNA-PKcs) forming the DNA-PK holoenzyme. DNA-PKcs recruitment occurs predominantly via the unique Ku80 C-terminal region, with holoenzyme formation activating DNA-PK activity (Gell & Jackson, 1999; Singleton, Torres-Arzayus, Rottinghaus, Taccioli, & Jeggo, 1999). *DNA-PKcs (PRKDC)* is an exceptionally large cDNA (12 Kb) encoding a serine/threonine protein kinase belonging to the phosphoinositol 3-kinase-like kinase family (Hartley et al., 1995). Prevailing evidence suggests that the major role of DNA-PK is to regulate NHEJ (Meek, Douglas, Cui, Ding, & Lees-Miller, 2007) and, although it can phosphorylate most NHEJ component proteins, DNA-PKcs autophosphorylation is the essential event for NHEJ (Chan & Lees-Miller, 1996; Cui et al., 2005; Douglas et al., 2007; Soubeyrand, Pope, Pakuts, & Hache, 2003). Two autophosphorylation clusters have been identified, designated ABCDE and PQR (Meek et al., 2007). Autophosphorylation is complex but appears to regulate three features: the magnitude of end processing, the inactivation of DNA-PK activity, and complex dissociation (Neal & Meek, 2011). Increasingly, controlled regulation of end processing is recognized as a critical step in controlling resection and the interface with HR, and DNA-PK is a strong contender for functioning in this regulation (Neal & Meek, 2011) (see below). Notwithstanding the enormous size of DNA-PKcs, considerable insight has been gained into its structure using a range of techniques. The structure encompasses multiple alpha-helical (HEAT) repeats (helix-turn-helix), which form a hollow, circular, and flexible structure (Sibanda, Chirgadze, & Blundell, 2010). The C-terminal kinase domain is located at the "head" of the structure, which also encompasses a small HEAT repeat domain. Importantly, the ring structure is not closed and has a gap at the base that potentially allows DNA-PKcs to encircle DNA. Thus, unlike Ku, which is a closed circular structure, DNA-PKcs possesses an opening to facilitate entry onto and release from DNA. Solution structures of DNA-PK (i.e., Ku and DNA-PKcs) with DNA have also been examined by small angle X-ray scattering providing insight into the dynamic changes that occur during DNA-PK activation and the process of trans-autophosphorylation (Hammel et al., 2010).

3.2.2 DSB end processing during NHEJ

Most endogenously generated DSBs, including those induced during V(D)J recombination, do not arise with $5'$ phosphate and $3'OH$ ends, the requisite end structures for polymerization and ligation, and therefore require processing prior to ligation. Such end processing can include removal of nonligatable ends, such as a 5'OH or 3'Ps, and of damaged bases and fill in by polymerases. Resection of DNA sequences at the junction ends represents a somewhat distinct form of end processing, which will be discussed in Section 6. Polynucleotide kinase 3'phosphatase (PNK or PNKP) is a bifunctional end-processing factor, which functions as a nucleotide kinase generating 3'P ends as well as a phosphatase removing 5'P ends (Bernstein et al., 2009). PNPK was known to function in SSB repair, and recent studies have demonstrating that it also has a critical end-processing role during NHEJ (Chappell, Hanakahi, Karimi-Busheri, Weinfeld, & West, 2002; Koch et al., 2004). Insight into its function at ss- and dsDNA ends has revealed a DNA-bending mechanism to facilitate substrate recognition (Garces, Pearl, & Oliver, 2011).

Interestingly, interactions between XRCC4 and PNKP have been reported to promote the interface between end processing and ligation during NHEJ (Mani et al., 2007). Additionally, PNKP is phosphorylated by DNA-PK/ATM enhancing its association with DSB sites as well as DSB repair (Segal-Raz et al., 2011; Zolner et al., 2011). The polymerase that functions during NHEJ has been analyzed in detail, with substantial evidence that Polymerase mu has a major function with additional evidence supporting an involvement of Polymerase lambda (Andrade, Martin, Juarez, Lopez de Saro, & Blanco, 2009; Chayot, Montagne, & Ricchetti, 2012; Lucas et al., 2009; Zucca et al., 2013).

During V(D)J recombination, hairpin-ended DSBs are generated by a transesterification mechanism via recombination activating gene 1 and 2 (RAG1/2) (Hiom & Gellert, 1997). Subsequent rejoining of these ends requires Artemis-dependent cleavage of the hairpin, a process that contributes to generating diversity at the junctions (Ma, Pannicke, Schwarz, & Lieber, 2002). Artemis also functions in processing a subset of IR-induced DSBs, although it is not clear if this role involves hairpin cleavage (Riballo et al., 2004). During V(D)J recombination, hairpin cleavage does not occur at the DSB apex, resulting in the requirement for fill in or deletion of nucleotides via processes involving terminal nucleotide transferase (TdT) and polymerase mu (Gozalbo-Lopez et al., 2009; Lewis, 1994).

3.2.3 DNA ligase IV, XRCC4, and XLF, the second NHEJ complex

DNA ligase IV is the NHEJ-specific ligase (Grawunder et al., 1997; Teo & Jackson, 1997; Wilson, Grawunder, & Lieber, 1997). The catalytic domain is located at the N-terminus, while the C-terminus encompasses two

BRCA1 carboxyl terminus (BRCT) domains separated by a short spacer region. DNA ligase IV interacts tightly with and depends on XRCC4 for stability (Riballo et al., 2009). Interaction with XRCC4 occurs via the spacer region between the two BRCT domains as well as the second BRCT domain (Sibanda et al., 2001). XRCC4 has a globular head and two alpha-helical-coiled coil regions, and interaction with DNA ligase IV occurs via a kink in the coiled coil region (Sibanda et al., 2001). XLF/Cernunnos represents a further component of the ligation complex and is structurally similar to XRCC4, although sharing limited sequence similarity (Ahnesorg, Smith, & Jackson, 2006; Buck et al., 2006; Li et al., 2008). However, XLF differs in having a less extended helical domain than XRCC4 and folds back upon itself, positioning the head domain toward the C-terminal region (Andres, Modesti, Tsai, Chu, & Junop, 2007; Li et al., 2008). XLF is less strongly bound to DNA ligase IV compared to the XRCC4:DNA ligase IV interaction and is dispensable for DNA ligase IV stability (Riballo et al., 2009). Several recent studies have suggested that XRCC4:XLF:DNA ligase IV assemble as a filament structure mediated by head to head interactions between XRCC4 and XLF, with the proposal that the filaments serve to bridge and align DNA ends to promote ligation (Andres et al., 2012; Hammel et al., 2011; Ropars et al., 2011). There is also evidence that DNA-PK facilitates the recruitment of the DNA ligase IV–XRCC4–XLF complex (Drouet et al., 2005; Mari et al., 2006).

Another recent factor that has been proposed to influence NHEJ is aprataxin and PNK-like factor (APLF). APLF together with PARP3 was shown to enhance NHEJ *in vitro* (Rulten et al., 2011). More recently, Ku80 was shown to recruit APLF to Ku–DNA complexes via a von Willebrand (vWa) domain in Ku80, promoting assembly of multiprotein Ku–DNA complexes and hence enhancing ligation (Grundy et al., 2012).

3.3. Additional end-joining processes

End-joining studies involving reporter constructs, either plasmid-based or genome integrated, have revealed the presence of end-joining pathway(s) that exploit short homology at the DNA junctions, the classification of which has been confusing (Mladenov & Iliakis, 2011). Available evidence suggests end-joining involving microhomology, defined as MMEJ, can arise by a mechanism dependent upon canonical NHEJ (c-NHEJ) proteins (Ku, DNA ligase IV) as well as by an alternative mechanism (termed Alt-NHEJ). Here, we have delineated c-NHEJ into a resection-independent process as well as CtIP and resection-dependent c-NHEJ (Patterson-Fortin, Shao, Bretscher, Messick, & Greenberg, 2010) (see below). The latter process could involve short microhomology regions revealed following resection to mediate the end-joining. A process of this nature has been shown to take place in vertebrate cells (Yun & Hiom, 2009).

There has been increasing focus on a distinct process of end-joining that does not exploit c-NHEJ factors, called Alternative NHEJ (Alt-NHEJ) (Bogue, Wang, Zhu, & Roth, 1997; Corneo et al., 2007; Iliakis, 2009; Wang, Perrault, Takeda, Qin, & Iliakis, 2003). This process most likely represents a variant of MMEJ and, somewhat confusedly, is often defined as being synonymous with MMEJ. However, since most MMEJ studies do not define the genetic components and there is evidence for a form of c-NHEJ dependent upon microhomology, we have defined MMEJ as representing any end-joining involving microhomology. Thus, Alt-NHEJ represents one of several processes exploiting microhomology and falling into the MMEJ classification. A potential model for resection-dependent c-NHEJ based on published findings and our own unpublished findings is shown in Figure 1.3. Figure 1.3 legend highlights the speculative aspects of this model.

Alt-NHEJ involves PARP1, XRCC1, and Ligase 1 or Ligase III, as well as potentially the MRN complex (Della-Maria et al., 2011; Deriano, Stracker, Baker, Petrini, & Roth, 2009; Iliakis, 2009; Simsek et al., 2011; Wang et al., 2005) (Figure 1.3). The process, therefore, appears to represent a form of two coupled SSB rejoining events potentially using microhomology to tether the ends. The process functions predominantly in the absence of Ku and does not appear to function in the presence of Ku but absence of DNA-PKcs (Difilippantonio et al., 2000; Simsek & Jasin, 2010; Wang et al., 2006). Interestingly, however, there is evidence that Alt-NHEJ contributes to rejoining in the absence of DNA ligase IV, raising the possibility that in the absence of DNA-PKcs, Ku remains bound at the ends precluding resection while in the presence of DNA-PKcs the failure to progress NHEJ could lead to resection and Alt-NHEJ. Interestingly, specific mutations in the RAG proteins can also favor the usage of Alt-NHEJ during V(D)J recombination (Corneo et al., 2007). Alt-NHEJ perhaps most clearly contributes to abnormal CSR events that occur during immune development in Ku or DNA ligase IV negative cells (Boboila, Alt, & Schwer, 2012; Boboila et al., 2010). It is possible that DSBs arising during CSR represent two closely linked SSBs and may be particularly amenable to this form of end-joining. The process has been demonstrated to arise additionally at

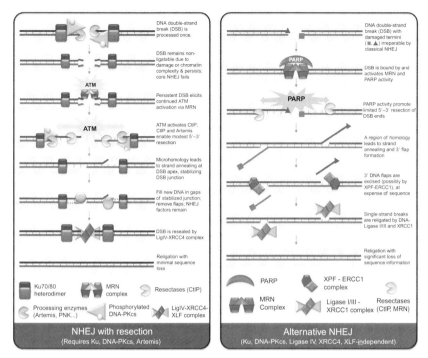

Figure 1.3 Additional mechanisms of DSB rejoining involving DNA end resection: Left panel, cyan. The analysis of DSB rejoining using plasmid substrates has suggested that a subprocess of NHEJ can arise following resection of the DNA ends. The precise details of this process are unknown and the figure depicts a likely model based on the available evidence. Following Ku and DNA-PKcs recruitment, NHEJ makes the initial attempt to rejoin a DSB. If rejoining does not ensue rapidly, then the ends can undergo CtIP-dependent resection generating short 3' overhangs. Resection is less efficient in G1 phase compared to S/G2 phase so that the magnitude of resection is small. This can then allow NHEJ using short regions of microhomology to aid synapsis. It is not currently known whether Ku vacates the DSB end to allow resection and subsequently rebinds, whether there is ongoing breathing of Ku at the ends or whether it translocates inter-nally. This may represent the slow process of DSB rejoining in G1 phase cells. Right panel, green: Alt-NHEJ predominantly takes place in cells lacking the Ku heterodimer, DNA ligase IV, or XRCC4. There is, however, additional evidence that it may represent a low-frequency process that causes translocation formation. The process involves the MRN complex, which is likely required to initiate resection. Any role for CtIP has not yet been shown. Alt-NHEJ also involves PARP1, which likely binds to the single-stranded DNA ends. Rejoining mainly involves DNA ligase III and XRCC1 with redundancy by DNA ligase I. We have speculated that XPF-ERCC1 is required for removal of the 3'DNA flaps. (For interpretation of the references to color in this figure legend, the reader is referred to the online version of this chapter.)

DSBs generated by zinc finger nucleases and, indeed, to contribute to translocation formation (Boboila et al., 2012; Simsek et al., 2011). There is also evidence that in certain tumor cells, c-NHEJ can be downregulated, leading to promotion of Alt-NHEJ (reviewed in Rassool & Tomkinson, 2010). Although Alt-NHEJ's significance requires further demonstration, there is provocative evidence that it represents a low-frequency rejoining event that nonetheless has the potential to make a significant contribution to translocation formation particularly those contributing to the etiology of lymphoid tumors (Difilippantonio et al., 2000; Nussenzweig & Nussenzweig, 2007; Yan et al., 2007).

4. DNA DAMAGE RESPONSE SIGNALING

In addition to repair mechanisms, DSBs activate a signal transduction process that drives a range of cellular consequences. Ataxia telangiectasia mutated (ATM) protein lies at the core of the DSB signaling response although in certain circumstances (e.g., following DSB end resection or if DSBs become encountered at the replication fork), ataxia telangiectasia mutated and Rad3 related (ATR) can also trigger a signaling-related cascade. Several outstanding reviews on the choreography of DSB signaling have been written (Ciccia & Elledge, 2010; Jackson & Bartek, 2009; Jeggo & Lavin, 2009; Panier & Durocher, 2009; Thompson, 2012). Only an overview with relevance to the ensuing discussion will be presented here.

4.1. The assembly process

The MRE11/RAD50/NBS1 (MRN) complex represents the initial DSB sensor, which via an interaction between ATM and NBS1 C-terminus, results in a first step tethering ATM at the DSB site (Ciccia & Elledge, 2010; Jackson & Bartek, 2009; Jeggo & Lavin, 2009; Panier & Durocher, 2009; Thompson, 2012). ATM then phosphorylates the histone variant, H2AX, generating γH2AX, which can extend megabase pair distances from the DSB and initiates the assembly of an array of DDR proteins that trigger further histone modifications in the DSB vicinity and influence downstream protein assembly at the DSB. In addition to phosphorylation, such modifications include ubiquitylation, sumoylation, and methylation (Al-Hakim et al., 2010; Gospodinov & Herceg, 2012; Polo & Jackson, 2011). The assembled proteins are called irradiation-induced foci (IRIF). Although the phosphorylation of downstream DDR proteins is ATM-dependent, γH2AX formation can be effected redundantly by DNA-PK or ATR

(Stiff et al., 2004). Thus, IRIF assembly can occur in the absence of ATM, although many aspects of the process are impaired. γH2AX formation facilitates the recruitment of MDC1, the first of several mediator proteins; MDC1 also interacts with MRN, which provides a second route to tether MRN (and hence ATM) at the DSB. MDC1 recruits two ubiquitin ligases, RNF8 and RNF168, to IRIF which ubiquitylate histone H2A in the DSB vicinity; there is increasing evidence that RNF8 may also ubiquitylate DDR proteins, such as Ku (Feng & Chen, 2012; Moyal et al., 2011). Histone H2A ubiquitylation results in the exposure or *de novo* formation of H4K20me3 (see below), which promotes 53BP1 recruitment (Mallette et al., 2012). 53BP1 also interacts with MRN and provides a further layer to tether ATM at the DSB site (Lee, Goodarzi, Jeggo, & Paull, 2010). Indeed, in the absence of 53BP1, ATM foci are substantially smaller.

In addition to this assembly cascade, another set of proteins accumulate at DSB sites either more substantially or exclusively in late S/G2 phase cells (Bekker-Jensen & Mailand, 2010). Central to this assembly is the recruitment of BRCA1, which requires RNF8-dependent ubiquitylation (Wang & Elledge, 2007). A complex of RAP80/BRCC36/ABRAXAS is also recruited possibly in a BRCA1-dependent manner (Shao et al., 2009) (see below).

4.2. Factors regulating the recruitment of 53BP1

There is emerging evidence that the accumulation of 53BP1 at DSB sites can block resection and RAD51 loading, and hence influence DSB repair pathway choice. Given the significance of 53BP1 in regulating pathway choice, we review current insight into the mechanism regulating 53BP1 positioning at DSBs. 53BP1's recruitment to IRIF requires its TUDOR domain, which promotes binding to dimethylated histones exposed in the DSB vicinity (Huyen et al., 2004). Histone H4K20me2 is the major factor required for 53BP1 chromatin binding (Botuyan et al., 2006). MMSET represents a histone methyl transferase that affects H4K20 methylation, and studies have suggested that it is recruited to DSBs and promotes 53BP1 binding (Pei et al., 2011). However, it is clear that the E3 ubiquitin ligases, RNF8 and RNF168, which promote the formation of K63-linked ubiquitin chains on H2A, are also critical for 53BP1 IRIF. Recent work provided important insight into this process, showing that two TUDOR domain-containing lysine demethylase enzymes, JMJD2A and JMJD2B, may compete with 53BP1 binding (Mallette et al., 2012). Significantly, JMJD2A/B undergoes polyubiquitination and degradation via an RNF8–RNF168-dependent process at DSB sites. Since JMJD2A/B can also bind to H4K20me2 (Kim et al., 2006; Lee, Thompson, Botuyan, & Mer, 2008), the model proposed is that JMJD2A/B regulate 53BP1 binding in a competitive manner (Mallette et al., 2012). Thus, overexpression of JMJD2A/B inhibits 53BP1 IRIF. Adding further complexity to this regulation is the proteasome factor AAA-ATPase p97/VCP, which has been suggested to promote loss of K48-linked ubiquitin chains and has been shown to have a partial impact on 53BP1 localization (Acs et al., 2011; Meerang et al., 2011). Additionally, p97/VCP regulates L3MBTL1, a distinct H4K20me2 binding protein (Acs et al., 2011); however, the interplay between L3MBTL1 and JMJD2A/B in regulating 53BP1 binding is unclear.

In summary, the above findings support the notion that 53BP1 binding is regulated by competitive binding of either (or both) L3MBTL1 or JMJD2A/B to H4K20me3. Upstream of these events, a further study reported that the proteasomal deubiquitinating enzyme POH1 influences the binding of JMJD2A to chromatin by regulating the level of RNF8/RNF169 K63-linked polyubiquitin chains (Butler et al., 2012). POH1 functions by removing the K63-linked ubiquitin chains and, thus, 53BP1 foci appear larger in the absence of POH1.

4.3. Assembly of the BRCA1 and the RAP80/Abraxas/BRCC36 complex

BRCA1's BRCT domain has been shown to regulate the formation of three distinct complexes (Kim, Chen, & Yu, 2007; Sobhian et al., 2007). The BRCA1-A complex encompasses RAP80, CCDC98/ABRAXAS, BRCC36, BRCC45/BRE, and MERIT40/NBA1 (Coleman & Greenberg, 2011; Feng, Wang, & Chen, 2010; Hu et al., 2011; Sobhian et al., 2007). Additional complexes involve BACH1 (BRIP1 or FANCJ) or CtIP, respectively. Here, we focus on the RAP80, ABRAXAS, and BRCC36 complex, which plays a significant role in regulating pathway choice (Coleman & Greenberg, 2011; Feng et al., 2010; Hu et al., 2011). RAP80 has a tandem ubiquitin-interaction motif allowing it to bind K6- or K63-linked ubiquitin chains (Messick & Greenberg, 2009). RAP80 has been reported to enhance BRCA1 recruitment to IRIF via this motif (Kim, Chen, et al., 2007; Sobhian et al., 2007; Wang & Elledge, 2007). However, although RAP80 depletion diminishes BRCA1 at IRIF, residual BRCA1 is recruited and, importantly, its function remains; indeed, while loss of BRCA1 results in markedly impaired resection and a failure to progress HR, loss of RAP80 leads to enhanced resection and hyper recombination (Hu et al., 2011). Further insight into this regulation was shown by dissection of the nature of the BRCA1–RAP80 complexes, showing that unbridled BRCA1-dependent end processing is caused by the loss of RAP80 (Coleman & Greenberg, 2011; Hu et al., 2011).

BRCC36 contains a JAMM (JAB/MPN/Mov34 metalloenzyme) domain and represents a K63Ub-specific deubiquitinating enzyme (DUB) (Feng et al., 2010; Patterson-Fortin et al., 2010; Shao et al., 2009; Sobhian et al., 2007). Abraxas and BRCC45 are essential for BRCC36 DUB activity within the RAP80 complex (Patterson-Fortin et al., 2010), and loss of BRCC36 results in elevated DSB end resection and enhanced HR (Coleman & Greenberg, 2011). Collectively, these findings suggest that the tight regulation of RAP80 binding to ubiquitin chains via the two DUBs, BRCC36 and BRCC45, could regulate BRCA1-dependent resection. Given the role of BRCA1 in regulating 53BP1, it is interesting to speculate that RAP80 could be involved in regulating 53BP1 localization at IRIF.

4.4. Involvement of noncoding RNAs

A very recent study provided unexpected insight into an early stage in the DDR activation (Francia et al., 2012). Although there is increasing recognition that noncoding RNAs (ncRNAs) control many aspects of cell responses, a role in the DDR had not previously been reported. ncRNA processing occurs via two RNases, DICER and DROSHA, which generate small dsRNAs that function in RNA interference. DICER and DROSHA (but not downstream components of the RNAi machinery) were required for efficient activation of the DDR via the generation of site-specific small RNAs, which activated the DDR in an MRN-dependent manner. The requirement of MRN for ATM activation strongly suggests a role for DICER and DROSHA in this process.

5. FUNCTIONS OF THE DDR ASSEMBLY

Ataxia telangiectasia (A-T), the human disorder caused by ATM mutation, is one of the most radiosensitive human conditions, demonstrating the importance of ATM signaling to the DSB response (Jeggo & Lavin, 2009). Studies of chromosomal breakage in A-T cells revealed increased persisting chromosome breaks (Jeggo & Lavin, 2009). However, DSB repair analysis assessed using procedures such as pulsed field gel electrophoresis

(PFGE) or the rate of loss of the DSB marker, γH2AX, revealed nearly normal repair kinetics, although 10–20% of DSBs do persist (Foray et al., 1997; Riballo et al., 2004).

The ATM signaling cascade is frequently described as representing the assembly of DSB repair factors. However, although the DDR proteins enhance DSB repair in subtle ways, they are not essential for NHEJ (Riballo et al., 2004). Indeed, most NHEJ occurs independently of ATM signaling and *vice versa*, ATM signaling functions independently of the assembly of NHEJ proteins. Indeed, the two processes appear to be relatively noninterfacing. Nonetheless, ATM signaling is essential for HR at DSBs and regulates other aspects of DSB repair. Likewise, the signaling response has impacts distinct to DSB repair. We consider three impacts of DDR assembly. Firstly, DDR assembly enhances ATM activation and its communication with the cell cycle checkpoint and/or apoptotic machinery, activating a plethora of transcriptional changes, which have wide-ranging impacts. Second, the assembled proteins include numerous chromatin remodeling factors that affect changes to the chromatin in the DSB vicinity. Such changes likely include compaction at certain DSBs and relaxation at others. These changes can influence the repair process as well as the magnitude and duration of ATM signaling. Finally, a related but somewhat distinct endpoint of ATM signaling is the regulation of the interplay between NHEJ and HR. Given the intense study of this aspect of DSB repair in recent years, it will be considered in a distinct section below.

5.1. Signaling to the checkpoint machinery

The induction of cell cycle checkpoint arrest, apoptosis, and transcriptional changes are highly important end-points of DDR signaling (Bartek & Lukas, 2007). Although ATM is essential for checkpoint arrest, a subfraction of apoptosis and most transcriptional changes after IR, the requirement for other DDR components is generally less marked (Fernandez-Capetillo et al., 2002). Indeed, H2AX as well as MDC1 and 53BP1 are not essential for G2/M checkpoint arrest, except after exposure to low-radiation doses. The proposed explanation is that the mediator proteins (and γH2AX formation) amplify but are not essential for ATM activation at DSBs. This can influence checkpoint arrest after exposure to low doses where a threshold level of signaling is not reached, while at high doses a level of signaling sufficient to activate checkpoint arrest can be attained without the mediator proteins (Deckbar et al., 2007; Lobrich & Jeggo, 2007). Interestingly, the duration of checkpoint arrest after exposure to high doses is compromised when the mediator proteins are lost, consistent with the notion that they amplify the checkpoint signal as DSB numbers decline following repair (Lobrich & Jeggo, 2007; Shibata et al., 2010). Whether the apoptosis or transcription changes are similarly influenced has not been examined rigorously.

Cell cycle checkpoint arrest has been reviewed previously and will not be described in detail here (Bartek & Lukas, 2003, 2007; Deckbar, Jeggo, & Lobrich, 2011; Wahl, Linke, Paulson, & Huang, 1997). We, however, consider the cellular impact of checkpoint arrest. Three significant checkpoints have been unearthed: G2/M, G1/S, and intra-S phase arrest. G2/M checkpoint arrest involves the phosphorylation and activation of the CHK1/CHK2 transducer kinases which, via their phosphorylation of the CDC25 phosphatases and subsequent regulation of CDK1 activity, causes rapid inhibition of mitotic entry (Bartek & Lukas, 2003, 2007). Significantly, the process has a defined threshold of sensitivity with both the activation and maintenance of G2/M arrest failing when the number of DSBs are low, with estimates suggesting that 10–15 DSBs are required for efficient checkpoint activation (Deckbar et al., 2007, 2011; Fernet, Megnin-Chanet, Hall, & Favaudon, 2009; Shibata et al., 2009). Consequently, the G2/M checkpoint fails to fully preclude the formation of chromosomal DSBs in mitotic cells and/or the formation of translocations. Nonetheless, G2/M arrest makes a significant contribution to the maintenance of genomic stability by enhancing the time available for DSB repair. This may be particularly important for HR in G2, a slow DSB repair process (Lobrich & Jeggo, 2007). Although G2/M arrest may not contribute dramatically to survival post IR, it does underlie the phenomenon of low-dose hypersensitivity (Marples & Collis, 2008).

Two processes have been reported to contribute to G1/S checkpoint arrest (Bartek & Lukas, 2001; Wahl et al., 1997): the first slows but does not fully prevent S phase entry and is an analogous process to G2/M arrest involving CHK1/2-dependent inhibition of CDK (CDK2 for G1/S arrest compare to CDK1 for G2/M arrest) (Deckbar et al., 2010). The second, more widely studied and significant process, represents ATM (after IR) and p53-dependent activation of the CDK inhibitor, p21 (Kastan et al., 1992; Wahl et al., 1997). This process requires transcriptional activation and thus only inhibits S phase entry several hours post IR (Deckbar et al., 2010). Nonetheless, it plays a major role in maintaining genomic stability as demonstrated by the frequent loss of p53 during tumorigenesis

(Kastan & Bartek, 2004) and as a "second tier" mechanism to the inefficient G2/M checkpoint. The G1/S checkpoint is activated by low numbers—possibly just one—of DSBs (Deckbar et al., 2010). Intra-S phase checkpoint arrest is important to prevent replication progression in the presence of DSBs. Similar to G1/S arrest, there are two components to this process, involving the inhibition of ongoing replication fork progression and the inhibition of late firing origins.

5.2. The repair of DSB located within heterochromatin

Analysis of chromosomal DSB repair using premature chromosome condensation in nonreplicating G0 cells provided strong evidence that ATM impacts upon the DSB repair (Cornforth & Bedford, 1985; Jeggo & Lavin, 2009). However, the mechanistic basis underlying this finding remained unclear for many years. Finally, major insight into the role of ATM in DSB repair came from exploiting the enumeration of γH2AX foci to monitor DSB repair after low, physiologically relevant doses of radiation, where a subtle defect was observed (Riballo et al., 2004). Measurement of the kinetics of DSB repair using either physical methods (e.g., PFGE) or γH2AX analysis had revealed the presence of a fast process, which repairs approximately 80% of radiation-induced DSBs, and a slower process repairing the remaining 20% of DSBs (Iliakis, Metzger, Denko, & Stamato, 1991; Riballo et al., 2004). Strikingly, ATM is required specifically for the slow DSB repair process (Riballo et al., 2004). In G1 phase, this process also requires Artemis and the DDR signaling proteins, H2AX, MDC1, and 53BP1 (Riballo et al., 2004). Indeed, the process necessitates the recruitment of 53BP1 to DSB sites and hence requires all DDR signaling proteins needed to recruit 53BP1 (Noon et al., 2010). Of note, in G2 phase, the slow component of DSB repair represents HR (Beucher et al., 2009) (see below). One explanation for the slow DSB repair process was that it represented the repair of more complex DSBs induced by IR (Riballo et al., 2004). However, subsequent studies demonstrated that there was a similar magnitude of requirement for ATM and the DDR signaling proteins following exposure to neocarzinostatin or calicheamicin, agents which induce more homogenous and less complex DSBs (Goodarzi et al., 2008). With increasing evidence that highly compacted heterochromatin (HC) influences DDR signaling and the fact that ~20% of the human genome is compacted into HC, studies were undertaken to examine whether the DSBs that require ATM for repair could represent those located in HC (designated

HC-Dsbs) (Cowell et al., 2007; Falk, Lukasova, & Kozubek, 2008; Kim, Kruhlak, Dotiwala, Nussenzweig, & Haber). The use of NIH3T3 cells, where HC regions can be readily visualized by their dense DAPI staining, revealed that the DSBs remaining in the absence of ATM localize to the periphery of HC regions (Goodarzi et al., 2008). Notably, the HC building factor KAP-1 is robustly phosphorylated at S824 by ATM after DSB formation (Ziv et al., 2006). Significantly, phosphorylated KAP-1 (pKAP-1) can arise in a low intensity (but dose dependent) pan-nuclear manner but additionally forms discrete pKAP-1 foci; the latter forming uniquely at HC-DSBs (Noon et al., 2010). Although both processes are ATM dependent, only the latter requires 53BP1. Based on an interaction between 53BP1 and RAD50, a component of the MRN complex, it was proposed that 53BP1 aids the tethering of ATM at DSBs and that this is required to promote robust pKAP-1 (i.e., pKAP-1 foci) (Lee et al., 2010; Noon et al., 2010). Given the requirement for 53BP1 for pKAP-1 foci formation at HC-DSBs and for HC-DSB repair, it was proposed that robust KAP-1 phosphorylation at HC-DSBs is required to achieve sufficient HC relaxation to overcome HC's negative impact on DSB repair. Consistent with this model, siRNA of several HC components or expression of the phosphomimetic KAP-1^{S824E} relieved the need for ATM for HC-DSB repair. KAP-1 undergoes auto-SUMOylation (Ivanov et al., 2007) and interacts with SUMO-interaction motifs (SIMs) present on factors required for heterochromatinization including the larger isoform of the nucleosome remodeling and deacetylase (NURD) complex component, the chromodomain helicase DNA-binding protein 3 (CHD3) (Ivanov et al., 2007; Schultz, Friedman, & Rauscher, 2001). Providing insight into the mechanism by which pKAP-1 affects the HC superstructure, it was recently proposed that a unique ATM substrate motif (824) on KAP-1, once phosphorylated, can outcompete the SUMO-SIM interaction between SUMOylated KAP-1 and the larger isoform of CHD3, resulting in CHD3 dispersal from HC regions in the vicinity of the DSB (Goodarzi, Kurka, & Jeggo, 2011). This represents a rapid and elegant way to achieve transient and localized HC relaxation without affecting epigenetic marks in the DSB vicinity. Collectively, these findings suggest that HC is a barrier to DSB repair and that ATM signaling serves to relieve that barrier. Given that only a small number of unrepaired DSBs can dramatically enhance the sensitivity to radiation, this provided at least, in part, an explanation for the marked radiosensitivity of A-T cells.

Yeast studies have provided strong evidence that HC regions are inhibitory to γH2AX foci spreading (Kim, Kruhlak, et al., 2007). In mammalian cells, γH2AX foci formation is not observed within the central mass of HC-rich chromocenters, rather forming only on their periphery (Chiolo et al., 2011; Falk, Lukasova, & Kozubek, 2010; Goodarzi et al., 2008). More recent studies have shown that DSBs induced within HC rapidly relocalize to the HC-periphery, potentially due to the inherent relaxation they trigger (Jakob et al., 2011). Thus, failure to observe γH2AX foci within HC regions may simply be because DSBs within HC are relocalized. However, γH2AX foci at HC regions are enlarged, with greater overlap between γH2AX and HC in cell lines from patients with disordered HC or following siRNA-induced HC relaxation, suggesting that HC serves as a barrier to IRIF expansion; such lines also show a diminished requirement for ATM during DSB repair (Brunton et al., 2011). An important, additional observation is that cell lines from such patients display hypersensitive initiation and maintenance of checkpoint arrest (Brunton et al., 2011). Collectively, this provides strong evidence that ATM signaling facilitates both repair and signaling to the checkpoint machinery at HC-DSBs.

5.3. Transcriptional arrest at transcriptionally active DSBs

Substantial evidence has shown that ultraviolet light (UV)-induced photoproducts pose a physical barrier to replication and transcription and that cells exploit specific mechanisms to counter this threat. Nucleotide excision repair (NER) removes UV photoproducts, while transcription-coupled repair, a modified form of NER, recognizes RNA pol II stalled at bulky lesions. When stalling is prolonged, monoubiquitylation of the stalled RNA pol II targets it for proteasome-mediated degradation (Harreman et al., 2009; Malik, Bagla, Chaurasia, Duan, & Bhaumik, 2008). Consequently, there is a decrease in transcription after UV exposure. For many years, searches for a similar process in response to DSB formation yielded no apparent difference in overall transcription rates, even following large doses. However, given the dramatic changes that occur in chromatin architecture in the DSB vicinity, including the recruitment of chromatin silencing factors, it is likely that transcription will be affected. Recently, a system using the Fok1 endonuclease to generate DSBs at lac operator repeats located distally to a promoter driving the transcription of a reporter gene was established to examine transcriptional repression at targeted DSBs (Shanbhag, Rafalska-Metcalf, Balane-Bolivar, Janicki, & Greenberg, 2010).

Transcription was regulated by tandem tetracycline response elements that bind a doxycycline-inducible promoter and monitored by fluorescent protein expression; hence, transcriptional activation or silencing could be monitored at a single cell level. The introduction of DSBs within 4 Kb of the transcribed gene caused transcriptional silencing, via a process requiring ATM and (partially) the RNF8 and RNF168 ubiquitin ligases, but not 53BP1. This suggests that an ATM-dependent process prevents RNA pol II elongation via chromatin decondensation at regions distal to DSBs.

Using genome wide, high-resolution mapping approaches, it has also been shown that the spreading of H2AX phosphorylation does not occur uniformly throughout the genome but rather that γH2AX-devoid or reduced regions can be observed at RNA pol II enriched sites undergoing active transcription (Iacovoni et al., 2010). Thus, there could be mechanisms to prevent γH2AX formation in transcribing genes. A more recent study from the same group has provided evidence that SMC3, a component of the cohesion complex, is recruited to DSBs and antagonizes γH2AX distribution within specific domains. Depletion of cohesin was observed to increase γH2AX spreading in specific domains diminished protein expression. However, whether these changes represent a consequence of altered chromatin caused by loss of cohesion or a specific impact of cohesin at DSBs remains unclear (Caron et al., 2012). Somewhat distinct to the study using multiple Fok1-induced DSBs, the impact of DSBs on transcription was also examined using the I-PpoI meganuclease to induce a single DSB in an RNA pol II transcribed gene (Pankotai, Bonhomme, Chen, & Soutoglou, 2012). In this case, inhibition of transcription elongation was observed to be DNA-PKcs-dependent and independent of ATM. RNA pol II was evicted from the coding and promoter region by a process involving proteasome-dependent degradation. The basis underlying the difference between these two studies remains unclear. It could be a consequence of the distinct number of DSBs induced, but additionally I-PpoI induces distinct DSBs to those induced by the Fok1 nuclease. Further, I-PpoI sites tend to lie within transcriptionally active rDNA genes, which are in close proximity within HC regions.

A further study has reported the inhibition of RNA pol I-dependent transcription at rDNA genes (Kruhlak et al., 2007), distinct to the modulation of RNA pol II-dependent transcription described earlier. RNA pol I-dependent transcription arrest is ATM and MDC1-dependent, partially NBS1-dependent and -independent of H2AX and 53BP1. Similar to the above processes, it represents a process targeting transcription in the DSB vicinity rather than global transcriptional arrest. Although further studies are required to unearth the interplay between transcription and DSB repair, clearly processes function to alert the transcription machinery to the presence of DSBs.

5.4. Impact of additional chromatin changes at DSBs
5.4.1 Role for HP1 in the DDR
Heterochromatin protein 1 (HP1) appears to have two opposing impacts on the DDR. As discussed earlier, by promoting heterochromatinization, HP1 suppresses DSB repair, necessitating ATM to effect localized HC relaxation (Goodarzi et al., 2008). Consistent with the notion that HC also suppresses ATM signaling, hyperactive G2/M checkpoint arrest is observed after HP1 siRNA (Kristin Zimmerman, unpublished findings). However, these impacts of HP1 represent a consequence of its role in chromatin organization and not a direct role in the DDR. Indeed, no changes in HP1 levels at HC-DSBs at later time points post-IR (when HC-DSB repair is taking place) have been observed (Goodarzi et al., 2008). However, more direct roles for HP1 in the DDR have been reported. One study observed that CK2-dependent HP1 phosphorylation occurred at early times after DNA damage, which results in transient displacement of HP1β from IRIF (Ayoub, Jeyasekharan, Bernal, & Venkitaraman, 2008; Ayoub, Jeyasekharan, & Venkitaraman, 2009). However, in other studies, transient recruitment of all three HP1 isoforms (α, B and γ) to DSBs was observed (Baldeyron, Soria, Roche, Cook, & Almouzni, 2011; Luijsterburg & van Attikum, 2011). One study exploited photobleaching at DSB sites to monitor the mobility of HP1 and observed that a small fraction of HP1 became immobile, suggesting increased chromatin binding or retention. This process required the chromoshadow domain of HP1α and was independent of H3K9me3, which did not change at the DSB site (Luijsterburg & van Attikum, 2011). A later study observed transient recruitment of HP1α at DNA damage induced by laser tracks, with the recruitment of HP1 dependent upon the PxVxL domain of p150 chromatin assembly factor 1 (CAF-1), the large subunit of CAF-1 (Baldeyron et al., 2011). Additionally, p150 CAF-1 siRNA impaired the recruitment of 53BP1, BRCA1, and RAD51 to IRIF and, most significantly, conferred a defect in HR. Further, KAP-1 was also transiently recruited to laser tracks.

5.4.2 The involvement of CHD4-NuRD in the DDR

As discussed earlier, CHD3/Mi2α, a subunit of NuRD, is dispersed from HC-DSBs during the DDR (Goodarzi et al., 2011). In contrast, CHD4/ Mi2β, another subunit of NuRD (present in a manner mutually exclusive with CHD3), has been reported to be recruited to DSB sites. CHD4 is a target of ATM/ATR and is recruited to laser tracks via a PARP-dependent process (Polo, Kaidi, Baskcomb, Galanty, & Jackson, 2010). Loss of CHD4 conferred reduced DSB repair and a failure to undergo G1/S checkpoint arrest. Surprisingly, however, neither DSB repair nor G1/S checkpoint arrest are PARP1 dependent. A more recent study also observed recruitment of CHD4 to laser tracks and its requirement for optimal stimulation of RNF8/RNF168-dependent ubiquitylation and hence BRCA1 recruitment, DSB repair, and checkpoint arrest (Larsen et al., 2010). CHD4 was later demonstrated to interact with RNF8 and mediate RNF8-dependent chromatin unfolding, independent of RNF8 Ub-ligase activity (Luijsterburg et al., 2012). The phosphorylation of CHD4 at S1349 by ATM is required for CHD4 retention at DSB sites, with S1349A mutants displaying reduced chromatin association; notably, CHD4 depletion did not impact overall ATM signaling but transiently increased persistent γH2AX levels up to, but not beyond, 6 h post IR (Urquhart, Gatei, Richard, & Khanna, 2011). Together, these findings provide provocative evidence for a role of CHD4 in the DDR, particularly relevant to repair events at early to intermediate time points, although its precise function requires further study.

5.4.3 The involvement of Tip60 and the NuA4 complex in the DDR

The Tip60 histone acetyltransferase (HAT) was first implicated in the DDR when it was shown to be required for acetylation and activation of ATM (Sun, Jiang, Chen, Fernandes, & Price, 2005). It was subsequently shown that Tip60 together with the HAT cofactor, Trrap, stimulates the acetylation of histone H4 at DNA damage sites and is required for HR; contrary to earlier work, ATM signaling under these conditions was found to be normal (Murr et al., 2006). In contrast, Tip60 was later found to be required for stimulating ATM activity following damage by promoting ATM monomerization via a process requiring ATM acetylation at K3016 (Sun, Xu, Roy, & Price, 2007). Both Tip60 and Trrap are components of the NuA4 chromatin remodeling complex alongside the SWI/SNF-type ATP-dependent chromatin remodeling enzyme, p400. All three components have now been implicated in DSB repair and DSB-induced chromatin modifications. Tip60, through its chromodomain, interacts directly with H3K9me3 exposed by the release of HP1 (as discussed above), and this event is dependent upon the MRN complex and is required for Tip60 activity (Sun et al., 2009). Together with the chromatin remodeling activity of p400, Tip60 is required for the chromatin alterations that are required for RNF8-dependent H2A ubiquitination (Xu et al., 2010). Together, these findings suggest that the NuA4 complex is required at multiple stages for a normal ATM signaling response.

5.4.4 The involvement of SNF2H, ACF1, and RNF20–RNF40 in the DDR

SNF2H, an ISWI-class ATP-dependent chromatin remodeling protein, is selectively recruited to DSBs in complex with the ACF1 protein (this complex is collectively referred to as ACF). ACF recruitment to DSBs appears to require the activity of the RNF20–RNF40 ubiquitin ligase heterodimer, which promotes ubiquitylation of histone H2B at K120 at sites of damage in an ATM-dependent manner following DSB induction (Lan et al., 2010; Moyal et al., 2011; Nakamura et al., 2011). RNF20 depletion or H2BK120 mutation confers significant defects to NHEJ and HR in plasmid assays, as well as reducing both Rad51 foci formation and laser microirradiation-induced XRCC4 recruitment to DSB tracks. Depletion of SNF2H or ACF1 showed similar impacts on NHEJ or HR efficiency. These findings provide strong evidence for SNF2H–ACF1 and RNF20–RNF40 playing key roles in both the primary DSB repair pathways, suggestive of a broad role in DNA end-joining (Figure 1.4). Further research will be needed to elucidate the molecular mechanism by which they function, particularly to define which (if any) chromatin alterations are regulated by their recruitment to DSBs and how these enable repair.

6. REGULATION OF DSB REPAIR PATHWAY CHOICE

6.1. Impact of cell cycle phase and resection

There is increasing evidence that the choice between DSB repair pathways is highly regulated and represents a significant function of IRIF assembly. Since HR is argued to be a more accurate DSB repair process and functions only in late S/G2, it was widely assumed that HR would represent the major DSB repair pathway in G2 phase (see below). However, studies focusing on the analysis of irradiated G2 cells (and inhibition of the progression of irradiated S phase cells into G2) have shown that, in fact, NHEJ is the

Chromatin alterations during the DNA double-strand break response

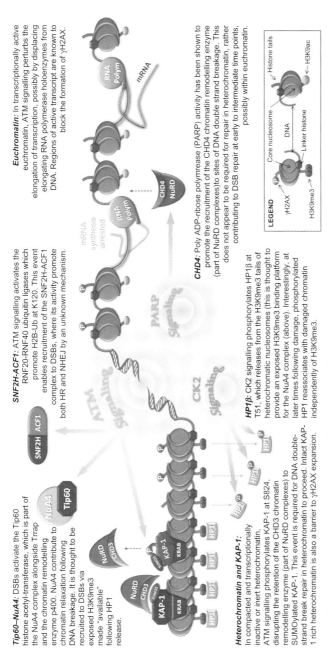

Tip60–NuA4: DSBs activate the Tip60 histone acetyl-transferase, which is part of the NuA4 complex alongside the Trrap enzyme and the chromatin remodelling enzyme p400. NuA4 contribute to chromatin relaxation following DNA breakage. It is thought to be recruited to DSBs via exposed H3K9me3 made "available" following HP1 release.

SNF2H–ACF1: ATM signalling activates the RNF20-RNF40 ubiquitin ligases which promote H2B-Ub at K120. This event enables recruitment of the SNF2H-ACF1 complex to DSBs, where its activity promote both HR and NHEJ by an unknown mechanism.

Euchromatin: In transcriptionally active euchromatin, ATM signalling perturbs the elongation of transcription, possibly by displacing elongating RNA polymerase holoenzymes from DNA. Regions of active transcript are known to block the formation of γH2AX.

Heterochromatin and KAP-1:
In compacted and transcriptionally inactive or inert heterochromatin, ATM signalling phosphorylates KAP-1 at S824, disrupting the retention of the CHD3 chromatin remodelling enzyme (part of NuRD complexes) to SUMOylated KAP-1. This event is required for DNA double-strand break repair in heterochromatin to proceed. Intact KAP-1 rich heterochromatin is also a barrier to γH2AX expansion.

HP1β: CK2 signalling phosphorylates HP1β at T51, which releases from the H3K9me3 tails of heterochromatic nucleosomes (this is thought to provide an exposed H3K9me3 binding platform for the NuA4 complex (above). Interestingly, at later times following damage, phosphorylated HP1 reassociates with damaged chromatin independently of H3K9me3.

CHD4: Poly ADP-ribose polymrease (PARP) activity has been shown to promote the recruitment of the CHD4 chromatin remodelling enzyme (part of NuRD complexes)to sites of DNA double strand breakage. This does not appear to be required for repair in heterochromatin, rather contributing to DSB repair at early to intermediate time points, possibly within euchromatin.

Figure 1.4 Chromatin alterations during the DSB response: DSB formation elicits numerous and dynamic changes to the architecture and ongoing processes within chromatin, each of which is specific to the type of chromatin within which the break occurs. Within heterochromatin (HC), ATM signalling mediates the release of CHD3-associated NuRD complex activity, by phosphorylating KAP-1 and disrupting its ability to retain CHD3 in HC via SUMO:SIM interactions. CK2 signaling triggers the release of HP1beta from H3K9me3 tails which may then be bound by ATM-activated NuA4 complexes containing Tip60. Collectively, these events, together with the ATM-dependent activation of RNF20/40 and consequent retention of SNF2H–ACF1 on chromatin surrounding the DSB site, are believed to effect changes necessary to enable repair by either NHEJ or HR to take place. In euchromatin, ATM signaling is known to perturb ongoing transcription and arrest mRNA synthesis. PARP activity promotes the activation and recruitment of CHD4-associated NuRD, potentially to effect chromatin changes needed for repair to take place efficiently. (For color version of this figure, the reader is referred to the online version of this chapter.)

predominant DSB repair pathway in G2 as in G1 (Beucher et al., 2009). In G1 and G2, DSB repair occurs with two component kinetics with the fast process representing NHEJ (Beucher et al., 2009; Riballo et al., 2004). However, specifically in G2 phase and not in G1, the slow process occurs by HR. An important initiating event for HR is DSB end resection, which can be visualized by the formation of RPA foci. Since Ku is highly abundant in mammalian cells and has avid dsDNA-binding capacity, the fact that NHEJ repairs most DSBs suggests that it represents the repair pathway of first choice. This notion has been supported by the finding that expression of a DNA-PKcs mutant which precludes autophosphorylation at the ABCDE cluster results in delayed RPA foci formation (Shibata et al., 2011). Previous studies had shown that the ABCDE nonphosphorylatable DNA-PKcs mutant shows delayed release from DSB ends (Hammel et al., 2010). The fact that a mutant form of DNA-PKcs could delay the onset of HR supports the notion that NHEJ functions upstream of HR in G2. Another important finding in this context is the observation that in G2 phase HR makes a greater contribution to DSB repair following irradiation with carbon ions, a form of high LET IR that induces highly complex DSB lesions with multiple damages in close proximity (Shibata et al., 2011). Studies in G1 phase have shown that carbon-ion-induced DSBs are repaired with slow kinetics due to the complexity of the DNA damage. Collectively, these studies support the model that NHEJ makes the first attempt to repair IR-induced DSBs but, if rapid repair does not ensue, then resection commits repair to HR. An additional observation is that the majority of DSBs that undergo repair by HR after X-irradiation of primary fibroblasts are those arising in HC (Beucher et al., 2009). Thus, it is argued that both DNA damage complexity and chromatin complexity can inhibit rejoining by NHEJ, allowing resection and a switch to HR.

6.2. Role of 53BP1

53BP1 is a key regulator of pathway choice, generally considered to promote NHEJ and restrict HR. The impact of 53BP1 siRNA on *I-Sce1* reporter assays shows an increased frequency of HR and reduced NHEJ in most studies consistent with the notion that 53BP1 suppresses HR (Xie et al., 2007). However, NHEJ assays do not represent simple end-ligation events, which would regenerate the I-Sce1 site, but frequently involve two closely localized DSBs. Thus, whether 53BP1 impacts upon NHEJ at simple two-ended DSBs remains unclear. An important observation revealing a role for 53BP1 was that 53BP1 loss in mice overcomes the genomic instability and HR defect conferred by BRCA1 loss, strongly suggesting that BRCA1 overrides the inhibitory impact of 53BP1 on HR, a notion consistent with the results using I-Sce1 assays (Bouwman et al., 2010). Further, 53BP1 forms as intense foci in G0/G1 phase cells but is excluded to the IRIF periphery in S/G2 cells via a BRCA1-dependent process, causing an overall reduction in the chromatin territory occupied by 53BP1 (Chapman, Sossick, Boulton, & Jackson, 2012). Taken together, these findings suggest that 53BP1 restricts resection and steps toward HR and that, to enable HR, it relocates in a BRCA1-dependent manner.

53BP1 exerts somewhat different roles in other cellular situations. Telomeres represent one-ended DSBs that are protected from the DSB repair machinery by the "shelterin" complex, a multisubunit complex that includes TRF2 and POT1 (Denchi & de Lange, 2007). In the absence of TRF2, uncapped telomeres activate ATM signaling and undergo repair by NHEJ generating telomere fusion events. While the loss of 53BP1 does not significantly impact telomere stability, combined loss of 53BP1 and TRF2 results in reduced telomere mobility (assessed by time-lapse microscopy) and reduced telomere fusion events (Dimitrova, Chen, Spector, & de Lange, 2008). It was proposed that 53BP1 affects chromatin changes influencing the ability of DNA ends to interact at distant sites.

V(D)J recombination involves the introduction of site-specific DSBs where two DSBs can either be close or distantly located but are held in a synaptonemal complex reported to involve 53BP1. Synapsis of long- but not short-range V(D)J recombination events appears to be 53BP1-dependent (Difilippantonio et al., 2008). In the absence of 53BP1, distal V(D)J recombination events undergo extensive degradation of coding ends and reintegration of signal joints at the original junctions. This defect is not observed in mice lacking other DDR proteins such as H2AX, MDC1, or ATM. This is perhaps related to the role of 53BP1 at telomere ends, with the suggestion that 53BP1's ability to oligomerize may facilitate synapsis enhancing the recombination or rejoining of distal ends.

Finally, 53BP1 is also required for CSR (Manis et al., 2004; Ward et al., 2004). CSR involves the introduction of site-specific DSBs by activation-induced cytidine deaminase (AID) (Muramatsu et al., 2000; Stavnezer, Guikema, & Schrader, 2008). AID is specifically activated in B cells undergoing CSR. AID, however, does not directly induce DSBs but rather promotes cytosine deamination to uracil, which is processed by the base excision repair machinery. Mutations at these sites can promote somatic hyper-mutation; although the process is not entirely understood, it appears that the generation of closely located SSBs via processing of closely located uracil residues can lead to the formation of DSBs (Staszewski et al., 2011). Rejoining of these DSBs appears to require the NHEJ proteins, but, unlike core NHEJ, the process additionally requires 53BP1 and the MRN complex (Manis et al., 2004; Ward et al., 2004). More recently, studies have suggested that 53BP1 is particularly important for CSR events that require rejoining of distal DNA ends, thereby relating to the role of 53BP1 in V(D)J recombination and telomere fusion. Further consolidating the models, it was suggested that 53BP1 can have two distinct impacts: (i) protecting DSBs from resection, thereby promoting core NHEJ and restricting homology mediated end-joining or recombination but (ii) promoting the rejoining of distant DSBs. The former process appears unrelated to distance but the latter functions optimally when DSBs are separated by ~100 Kb but not when the distance is as great as 2.7 Kb (Bothmer et al., 2011, 2010). These differing impacts of 53BP1 are depicted in Figure 1.5 and are discussed further in a previous review (Noon & Goodarzi, 2011).

In evaluating the roles of 53BP1, it is important to appreciate that the fast component of DSB repair after IR exposure, which represents NHEJ, occurs normally in cells lacking 53BP1. Indeed, diminished NHEJ is mainly observed in I-Sce1 assays where two closely localized DSB are induced. Clearly, 53BP1 is not essential for NHEJ and indeed may not impair NHEJ at normal two-ended DSBs but rather may serve to promote NHEJ when additional synapsis is required at more distally localized DSB ends. Further, 53BP1 has a role in tethering ATM at DSB ends to promote HC relaxation.

6.3. Model for the interface between NHEJ and HR

Collectively, the current data provide strong evidence that NHEJ makes the initial attempt to repair a DSB in G1 or G2 phase (Shibata et al., 2011). This is likely due to the abundance of Ku and its strong capacity to bind DSBs. It is likely that many DSBs are repaired rapidly by a process involving end processing but not necessarily end resection. Provided sequence information is not lost at the DSB junction, it is likely that NHEJ is an efficient DSB repair pathway. Indeed, it may be preferential to HR, which requires substantial chromatin modifications in the DSB vicinity since the recruitment of several factors to IRIF, such as HP1 and KAP1, appear to be specifically required for HR but are dispensable for NHEJ (Baldeyron et al., 2011). Ku binding suppresses resection of DNA ends (Symington & Gautier, 2011); however, if rapid rejoining does not occur due to the complexity of the DNA end or the chromatin environment, then resection can ensue

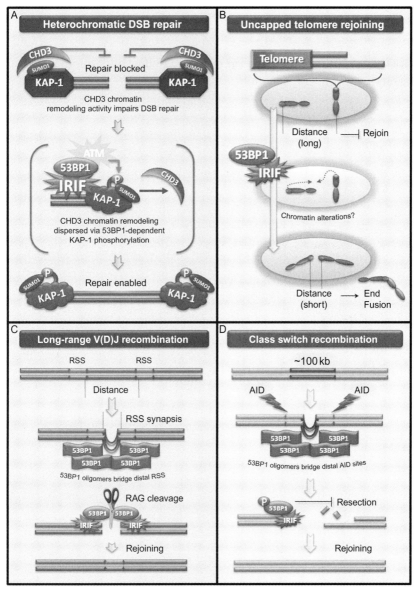

Figure 1.5 The impacts of 53BP1 on DNA DSB repair. (A) In heterochromatin, DSB repair is blocked by CHD3 ATP-dependent chromatin remodeling activity retained on SUMOylated KAP-1. 53BP1, once incorporated in IRIF, concentrates active ATM in the DSB vicinity to densely phosphorylate KAP-1, an event that disperses the larger isoform of CHD3 to enable repair. (B) Following deprotection of telomeres, these one-ended DSBs elicit a DDR in which 53BP1 assembles into IRIF. 53BP1 appears to increase the mobility of DSB ends (possibly via uncharacterized chromatin alterations), enabling chromosomal end-to-end fusion. (C) During V(D)J recombination of distally separated V, D, or J segments, recombination signal sequences (RSS) are brought in close proximity via oligomeric 53BP1, which maintains RSS synapsis until rejoining can take place. (D) 53BP1 oligomers promote rejoining of AID-induced DSB during CSR. Via the activity of glycosylases, sites deaminated by AID collapse into a DSB to which 53BP1 is recruited. 53BP1 supports the rejoining of DSBs separated by ~100 Kb. The phosphorylation of 53BP1's N-terminus is required for rejoining, although the mechanism by which this occurs is unclear.

in G2 phase. It is likely that resection in G2 requires the ejection of Ku from the DNA end, and, indeed, studies in yeast have suggested that MRN facilitates this (Symington & Gautier, 2011); this is less clear in mammalian cells. Nonetheless, there appears to be a regulated switch from NHEJ to a resection-dependent rejoining. 53BP1 and RAP80/BRCC36/ABRAXAS represent two components that serve as a barriers to resection (Coleman & Greenberg, 2011; Feng et al., 2010; Hu et al., 2011; Sobhian et al., 2007) with BRCA1 playing an important role to reposition 53BP1 to relieve its restriction to resection (Chapman et al., 2012).

The MRN/CtIP complex plays an important role in initiating resection (Symington & Gautier, 2011), and current models are consistent with a two-step process of resection with the initiating event being undertaken by the MRN/CtIP complex and extension of the end generated requiring Exo1/ BLM (Garcia et al., 2011). Findings strongly suggest that the initiation of resection by MRN/CtIP is the event committing to HR repair since CtIP allows repair to progress by NHEJ, whereas loss of the extension step of resection (e.g., by BRCA1 siRNA) causes a DSB repair defect (Shibata et al., 2011). Given this model, it is interesting to consider whether less marked resection might take place in G1 phase to allow resection-dependent NHEJ. Given the role of Artemis in the slow component of DSB repair in G1 phase and its role as a nuclease, we speculate that the slow component of DSB repair in G1 phase could, like HR in G2 phase, represent a form of resection-mediated DSB repair but involving less dramatic resection (Figure 1.1).

7. CONTRIBUTION OF DEFECTS IN DSB REJOINING PROCESSES TO HUMAN DISEASE

Given the role of NHEJ in V(D)J recombination, a striking and expected phenotype of mice lacking NHEJ proteins is severe combined immunodeficiency (SCID). Mutations in NHEJ proteins (DNA ligase IV, XLF, Artemis, and DNA-PKcs) have now been identified in a subclass of SCID or CID patients, defined as radiosensitive-SCID (RS-SCID) (O'Driscoll et al., 2001; O'Driscoll & Jeggo, 2006). Such patients undergo bone marrow transplantation (BMT) frequently, and the identification of such patients is important to avoid the use of DSB-inducing drugs during the BMT conditioning regime (O'Driscoll & Jeggo, 2008). Since Artemis is nonessential, mutations are frequently inactivating for Artemis function and patients present with SCID (Buck et al., 2006). Additional features associated with the loss of Artemis function are not severe and patients can be successfully treated with BMT. DNA ligase IV is essential and LIG4 Syndrome patients (deficient in DNA ligase IV) thus harbor hypomorphic mutational changes (O'Driscoll et al., 2001). Although the level of immunodeficiency observed in LIG4 syndrome and XLF-deficient patients is less severe than observed in Artemis-null patients, these two disorders are frequently associated with microcephaly and/or developmental delay, revealing the important role of NHEJ during neuronal development (O'Driscoll & Jeggo, 2006). Predisposition to lymphoid tumors is a further manifestation of these disorders, particularly in those patients with residual T and B cells, which likely represents the outcome of abnormal V(D)J recombination events. Prior to diagnosis, a single such patient who received radiotherapy to treat his leukemia died of radiation morbidity attesting to clinical radiosensitivity of these patients.

Since HR is an essential process, patients with mutations in these proteins are not commonly observed. However, heterozygous carriers with mutations in BRCA1 and BRCA2, two HR components, represent an important class of breast cancer predisposition patients, confirming the tumor suppressor function of HR. Importantly, mutations in several components of the cohesion complex, which promotes sister chromatid cohesion during HR, have been reported in disorders causing developmental abnormalities. Causal defects for Cornelia de Lange (CdLS) and Roberts Syndrome include *NIPBL, SMC1A*, and *SMC3*, and more recently, *HDAC8*, which encode proteins required for sister chromatid cohesion (reviewed in Horsfield, Print, & Monnich, 2012; Liu & Baynam, 2010). Interestingly, HDAC8, which is mutated in a small number of CdLS patients, was identified as the SMC3 deacetylase and defects caused inefficient dissolution of the cohesion complex during mitosis (Deardorff, Bando, et al., 2012). Mutations were also identified in RAD21, which encodes an integral cohesion structural protein, in a congenital phenotype with features including growth retardation, minor skeletal anomalies, and facial features overlapping with those of CdLS. Collectively, these disorders have been classified as cohesinopathy disorders (Deardorff, Wilde, et al., 2012).

Human disorders caused by mutations in the DDR signaling components have also been described. A-T, which causes a broad range of clinical features including progressive ataxia, telangiectasia, mild immunodeficiency, and cancer predisposition, is the best described of these disorders (Jeggo & Lavin, 2009); patients harboring mutations in all three components of the MRN complex (MRE11, RAD50, and NBS1) are also known

(Chrzanowska, Gregorek, Dembowska-Baginska, Kalina, & Digweed, 2012; Stewart et al., 1999; Waltes et al., 2009). Perhaps surprisingly, deficiency in MRE11 causes an Ataxia telangiectasia-like disorder while mutations in RAD50 or NBS1 cause distinct clinical features including microcephaly and developmental delay. Deficiency in RNF168 causes Riddle Syndrome, a disorder associated with radiosensitivity, immunodeficiency, dysmorphic facial features, and learning difficulties (Stewart et al., 2009). The features of these disorders have been described in detail previously (Blundred & Stewart, 2011; Chrzanowska et al., 2012).

8. CONCLUDING REMARKS

Here, we have overviewed the dramatic changes that arise as a consequence of DSB formation and consider how these changes impact upon the DSB repair process. Studies using lower organisms predicted that HR would represent the major DSB repair pathway in mammalian cells. However, it is increasingly evident that NHEJ carries out repair of most DSBs with HR functioning predominantly to handle lesions, including one-ended DSBs, at replication forks. One consideration underlying the greater significance of NHEJ in mammalian cells could be their increased level of chromatin complexity. Indeed, the dramatic chromatin changes that arise at a DSB are central to this review. While the basis underlying some of these changes is apparent already, the consequences of many of the changes are not yet fully appreciated and this represents an important area for future study. Additionally, HR may be downregulated in mammalian cells to avoid its usage with homologous chromosomes, where the level of homology may be insufficient to allow accurate repair. Nonetheless, HR does function, and a major consideration is how the interplay between these two processes is regulated. This likely represents an important component of the function of IRIF and further insight into this complex regulation is required. In this review, we have only briefly considered how these pathways and their regulation are changed in cancer cells. This represents an important area that is likely to emerge as highly significant during the coming years.

REFERENCES

Acs, K., Luijsterburg, M. S., Ackermann, L., Salomons, F. A., Hoppe, T., & Dantuma, N. P. (2011). The AAA-ATPase VCP/p97 promotes 53BP1 recruitment by removing L3MBTL1 from DNA double-strand breaks. *Nature Structural & Molecular Biology, 18*, 1345–1350.

Adamo, A., Collis, S. J., Adelman, C. A., Silva, N., Horejsi, Z., Ward, J. D., et al. (2010). Preventing nonhomologous end joining suppresses DNA repair defects of Fanconi anemia. *Molecular Cell, 39*, 25–35.

Ahnesorg, P., Smith, P., & Jackson, S. P. (2006). XLF interacts with the XRCC4-DNA ligase IV complex to promote DNA nonhomologous end-joining. *Cell, 124*, 301–313.

Al-Hakim, A., Escribano-Diaz, C., Landry, M. C., O'Donnell, L., Panier, S., Szilard, R. K., et al. (2010). The ubiquitous role of ubiquitin in the DNA damage response. *DNA Repair, 9*, 1229–1240.

Andrade, P., Martin, M. J., Juarez, R., Lopez de Saro, F., & Blanco, L. (2009). Limited terminal transferase in human DNA polymerase mu defines the required balance between accuracy and efficiency in NHEJ. [Research Support, Non-U.S. Gov't] *Proceedings of the National Academy of Sciences of the United States of America, 106*, 16203–16208.

Andres, S. N., Modesti, M., Tsai, C. J., Chu, G., & Junop, M. S. (2007). Crystal structure of human XLF: A twist in nonhomologous DNA end-joining. *Molecular Cell, 28*, 1093–1101.

Andres, S. N., Vergnes, A., Ristic, D., Wyman, C., Modesti, M., & Junop, M. (2012). A human XRCC4-XLF complex bridges DNA. *Nucleic Acids Research, 40*, 1868–1878.

Ayoub, N., Jeyasekharan, A. D., Bernal, J. A., & Venkitaraman, A. R. (2008). HP1-beta mobilization promotes chromatin changes that initiate the DNA damage response. *Nature, 453*, 682–686.

Ayoub, N., Jeyasekharan, A. D., & Venkitaraman, A. R. (2009). Mobilization and recruitment of HP1: A bimodal response to DNA breakage. *Cell Cycle, 8*, 2945–2950.

Baldeyron, C., Soria, G., Roche, D., Cook, A. J., & Almouzni, G. (2011). HP1alpha recruitment to DNA damage by p150CAF-1 promotes homologous recombination repair. *The Journal of Cell Biology, 193*, 81–95.

Bartek, J., & Lukas, J. (2001). Pathways governing G1/S transition and their response to DNA damage. *FEBS Letters, 490*, 117–122.

Bartek, J., & Lukas, J. (2003). Chk1 and Chk2 kinases in checkpoint control and cancer. *Cancer Cell, 3*, 421–429.

Bartek, J., & Lukas, J. (2007). DNA damage checkpoints: From initiation to recovery or adaptation. *Current Opinion in Cell Biology, 19*, 238–245.

Bekker-Jensen, S., & Mailand, N. (2010). Assembly and function of DNA double-strand break repair foci in mammalian cells. *DNA Repair (Amst), 9*, 1219–1228.

Bell, J. C., Plank, J. L., Dombrowski, C. C., & Kowalczykowski, S. C. (2012). Direct imaging of RecA nucleation and growth on single molecules of SSB-coated ssDNA. *Nature, 491*, 274–278.

Bernstein, N. K., Hammel, M., Mani, R. S., Weinfeld, M., Pelikan, M., Tainer, J. A., et al. (2009). Mechanism of DNA substrate recognition by the mammalian DNA repair enzyme, Polynucleotide Kinase. *Nucleic Acids Research, 37*, 6161–6173.

Beucher, A., Birraux, J., Tchouandong, L., Barton, O., Shibata, A., Conrad, S., et al. (2009). ATM and Artemis promote homologous recombination of radiation-induced DNA double-strand breaks in G2. *The EMBO Journal, 28*, 3413–3427.

Blow, J. J., Ge, X. Q., & Jackson, D. A. (2011). How dormant origins promote complete genome replication. *Trends in Biochemical Sciences, 36*, 405–414.

Blundred, R. M., & Stewart, G. S. (2011). DNA double-strand break repair, immunodeficiency and the RIDDLE syndrome. *Expert Review of Clinical Immunology, 7*, 169–185.

Boboila, C., Alt, F. W., & Schwer, B. (2012). Classical and alternative end-joining pathways for repair of lymphocyte-specific and general DNA double-strand breaks. *Advances in Immunology, 116*, 1–49.

Boboila, C., Jankovic, M., Yan, C. T., Wang, J. H., Wesemann, D. R., Zhang, T., et al. (2010). Alternative end-joining catalyzes robust IgH locus deletions and translocations in the combined absence of ligase 4 and Ku70. *Proceedings of the National Academy of Sciences of the United States of America, 107,* 3034–3039.

Bogue, M. A., Wang, C., Zhu, C., & Roth, D. B. (1997). V(D)J recombination in Ku86-deficient mice: Distinct effects on coding, signal, and hybrid joint formation. *Immunity, 7,* 37–47.

Borde, V., & de Massy, B. (2013). Programmed induction of DNA double strand breaks during meiosis: Setting up communication between DNA and the chromosome structure. *Current Opinion in Genetics & Development,* S0959–437X(12)00146–3.

Bothmer, A., Robbiani, D. F., Di Virgilio, M., Bunting, S. F., Klein, I. A., Feldhahn, N., et al. (2011). Regulation of DNA End Joining, Resection, and Immunoglobulin Class Switch Recombination by 53BP1. *Molecular Cell, 42,* 319–329.

Bothmer, A., Robbiani, D. F., Feldhahn, N., Gazumyan, A., Nussenzweig, A., & Nussenzweig, M. C. (2010). 53BP1 regulates DNA resection and the choice between classical and alternative end joining during class switch recombination. *The Journal of Experimental Medicine, 207,* 855–865.

Botuyan, M. V., Lee, J., Ward, I. M., Kim, J. E., Thompson, J. R., Chen, J., et al. (2006). Structural basis for the methylation state-specific recognition of histone H4-K20 by 53BP1 and Crb2 in DNA repair. *Cell, 127,* 1361–1373.

Bouwman, P., Aly, A., Escandell, J. M., Pieterse, M., Bartkova, J., van der Gulden, H., et al. (2010). 53BP1 loss rescues BRCA1 deficiency and is associated with triple-negative and BRCA-mutated breast cancers. *Nature Structural & Molecular Biology, 17,* 688–695.

Brunton, H., Goodarzi, A. A., Noon, A. T., Shrikhande, A., Hansen, R. S., Jeggo, P. A., et al. (2011). Analysis of human syndromes with disordered chromatin reveals the impact of heterochromatin on the efficacy of ATM-dependent G2/M checkpoint arrest. *Molecular and Cellular Biology, 31,* 4022–4035.

Buck, D., Malivert, L., de Chasseval, R., Barraud, A., Fondaneche, M. C., Sanal, O., et al. (2006). Cernunnos, a novel nonhomologous end-joining factor, is mutated in human immunodeficiency with microcephaly. *Cell, 124,* 287–299.

Butler, L. R., Densham, R. M., Jia, J., Garvin, A. J., Stone, H. R., Shah, V., et al. (2012). The proteosomal de Ubiquitinating enzyme POH1 promotes the double-strand DNA break response. *The EMBO Journal, 31,* 3918–3934.

Caron, P., Aymard, F., Iacovoni, J. S., Briois, S., Canitrot, Y., Bugler, B., et al. (2012). Cohesin protects genes against gammaH2AX Induced by DNA double-strand breaks. *PLoS Genetics, 8,* e1002460.

Carreira, A., & Kowalczykowski, S. C. (2011). Two classes of BRC repeats in BRCA2 promote RAD51 nucleoprotein filament function by distinct mechanisms. *Proceedings of the National Academy of Sciences of the United States of America, 108,* 10448–10453.

Chan, D. W., & Lees-Miller, S. P. (1996). The DNA-dependent protein kinase is inactivated by autophosphorylation of the catalytic subunit. *The Journal of Biological Chemistry, 271,* 8936–8941.

Chapman, J. R., Sossick, A. J., Boulton, S. J., & Jackson, S. P. (2012). BRCA1-associated exclusion of 53BP1 from DNA damage sites underlies temporal control of DNA repair. *Journal of Cell Science, 125,* 3529–3534.

Chappell, C., Hanakahi, L. A., Karimi-Busheri, F., Weinfeld, M., & West, S. C. (2002). Involvement of human polynucleotide kinase in double-strand break repair by non-homologous end joining. *The EMBO Journal, 21,* 2827–2832.

Chayot, R., Montagne, B., & Ricchetti, M. (2012). DNA polymerase mu is a global player in the repair of non-homologous end-joining substrates. [Research Support, Non-U.S. Gov't] *DNA Repair, 11,* 22–34.

Chiolo, I., Minoda, A., Colmenares, S. U., Polyzos, A., Costes, S. V., & Karpen, G. H. (2011). Double-strand breaks in heterochromatin move outside of a dynamic HP1a domain to complete recombinational repair. *Cell, 144,* 732–744.

Chrzanowska, K. H., Gregorek, H., Dembowska-Baginska, B., Kalina, M. A., & Digweed, M. (2012). Nijmegen breakage syndrome. *Orphanet Journal of Rare Diseases*, 7, 13.

Ciccia, A., & Elledge, S. J. (2010). The DNA damage response: Making it safe to play with knives. *Molecular Cell*, 40, 179–204.

Coleman, K. A., & Greenberg, R. A. (2011). The BRCA1-RAP80 complex regulates DNA repair mechanism utilization by restricting end resection. *The Journal of Biological Chemistry*, 286, 13669–13680.

Corneo, B., Wendland, R. L., Deriano, L., Cui, X., Klein, I. A., Wong, S. Y., et al. (2007). Rag mutations reveal robust alternative end joining. *Nature*, 449, 483–486.

Cornforth, M. N., & Bedford, J. S. (1985). On the nature of a defect in cells from individuals with ataxia-telangiectasia. *Science*, 227, 1589–1591.

Cowell, I. G., Sunter, N. J., Singh, P. B., Austin, C. A., Durkacz, B. W., & Tilby, M. J. (2007). gammaH2AX foci form preferentially in euchromatin after ionising-radiation. *PLoS One*, 2, e1057.

Cui, X., Yu, Y., Gupta, S., Cho, Y. M., Lees-Miller, S. P., & Meek, K. (2005). Autophosphorylation of DNA-dependent protein kinase regulates DNA end processing and may also alter double-strand break repair pathway choice. *Molecular and Cellular Biology*, 25, 10842–10852.

Deardorff, M. A., Bando, M., Nakato, R., Watrin, E., Itoh, T., Minamino, M., et al. (2012). HDAC8 mutations in Cornelia de Lange syndrome affect the cohesin acetylation cycle. [Research Support] *Nature*, 489, 313–317.

Deardorff, M. A., Wilde, J. J., Albrecht, M., Dickinson, E., Tennstedt, S., Braunholz, D., et al. (2012). RAD21 mutations cause a human cohesinopathy. *The American Journal of Human Genetics*, 90, 1014–1027.

Deckbar, D., Birraux, J., Krempler, A., Tchouandong, L., Beucher, A., Walker, S., et al. (2007). Chromosome breakage after G2 checkpoint release. *The Journal of Cell Biology*, 176, 748–755.

Deckbar, D., Jeggo, P. A., & Lobrich, M. (2011). Understanding the limitations of radiation-induced cell cycle checkpoints. *Critical Reviews in Biochemistry and Molecular Biology*, 46, 271–283.

Deckbar, D., Stiff, T., Koch, B., Reis, C., Lobrich, M., & Jeggo, P. A. (2010). The limitations of the G1-S checkpoint. *Cancer Research*, 70, 4412–4421.

Della-Maria, J., Zhou, Y., Tsai, M. S., Kuhnlein, J., Carney, J. P., Paull, T. T., et al. (2011). Human Mre11/human Rad50/Nbs1 and DNA ligase IIIalpha/XRCC1 protein complexes act together in an alternative nonhomologous end joining pathway. *The Journal of Biological Chemistry*, 286, 33845–33853.

Denchi, E. L., & de Lange, T. (2007). Protection of telomeres through independent control of ATM and ATR by TRF2 and POT1. *Nature*, 448, 1068–1071.

Deriano, L., Stracker, T. H., Baker, A., Petrini, J. H., & Roth, D. B. (2009). Roles for NBS1 in alternative nonhomologous end-joining of V(D)J recombination intermediates. *Molecular Cell*, 34, 13–25.

Difilippantonio, S., Gapud, E., Wong, N., Huang, C. Y., Mahowald, G., Chen, H. T., et al. (2008). 53BP1 facilitates long-range DNA end-joining during V(D)J recombination. *Nature*, 456, 529–533.

Difilippantonio, M. J., Zhu, J., Chen, H. T., Meffre, E., Nussenzweig, M. C., Max, E. E., et al. (2000). DNA repair protein Ku80 suppresses chromosomal aberrations and malignant transformation. *Nature*, 404, 510–514.

Dimitrova, N., Chen, Y. C., Spector, D. L., & de Lange, T. (2008). 53BP1 promotes non-homologous end joining of telomeres by increasing chromatin mobility. *Nature*, 456, 524–528.

Douglas, P., Cui, X., Block, W. D., Yu, Y., Gupta, S., Ding, Q., et al. (2007). The DNA-dependent protein kinase catalytic subunit is phosphorylated in vivo on threonine 3950, a highly conserved amino acid in the protein kinase domain. *Molecular and Cellular Biology*, *27*, 1581–1591.

Drouet, J., Delteil, C., Lefrancois, J., Concannon, P., Salles, B., & Calsou, P. (2005). DNA-dependent protein kinase and XRCC4-DNA ligase IV mobilization in the cell in response to DNA double strand breaks. *The Journal of Biological Chemistry*, *280*, 7060–7069.

Falk, M., Lukasova, E., & Kozubek, S. (2008). Chromatin structure influences the sensitivity of DNA to gamma-radiation. *Biochimica et Biophysica Acta*, *1783*, 2398–2414.

Falk, M., Lukasova, E., & Kozubek, S. (2010). Higher-order chromatin structure in DSB induction, repair and misrepair. *Mutation Research*, *704*, 88–100.

Feng, L., & Chen, J. (2012). The E3 ligase RNF8 regulates KU80 removal and NHEJ repair. *Nature Structural & Molecular Biology*, *19*, 201–206.

Feng, L., Wang, J., & Chen, J. (2010). The Lys63-specific deubiquitinating enzyme BRCC36 is regulated by two scaffold proteins localizing in different subcellular compartments. *The Journal of Biological Chemistry*, *285*, 30982–30988.

Fernandez-Capetillo, O., Chen, H. T., Celeste, A., Ward, I., Romanienko, P. J., Morales, J. C., et al. (2002). DNA damage-induced G2-M checkpoint activation by histone H2AX and 53BP1. *Nature Cell Biology*, *4*, 993–997.

Fernet, M., Megnin-Chanet, F., Hall, J., & Favaudon, V. (2009). Control of the G2/M checkpoints after exposure to low doses of ionising radiation: Implications for hyper-radiosensitivity. *DNA Repair (Amst)*, *9*, 48–57.

Foray, N., Priestley, A., Alsbeih, G., Badie, C., Capulas, E. P., Arlett, C. F., et al. (1997). Hypersensitivity of ataxia-telangiectasia fibroblasts to ionizing radiation is associated with a repair deficiency of DNA double-strand breaks. *International Journal of Radiation Biology*, *72*, 271–283.

Forget, A. L., & Kowalczykowski, S. C. (2012). Single-molecule imaging of DNA pairing by RecA reveals a three-dimensional homology search. *Nature*, *482*, 423–427.

Francia, S., Michelini, F., Saxena, A., Tang, D., de Hoon, M., Anelli, V., et al. (2012). Site-specific DICER and DROSHA RNA products control the DNA-damage response. [Research Support, Non-U.S. Gov't] *Nature*, *488*, 231–235.

Garces, F., Pearl, L. H., & Oliver, A. W. (2011). The structural basis for substrate recognition by mammalian polynucleotide kinase 3' phosphatase. *Molecular Cell*, *44*, 385–396.

Garcia, V., Phelps, S. E., Gray, S., & Neale, M. J. (2011). Bidirectional resection of DNA double-strand breaks by Mre11 and Exo1. [Research Support, Non-U.S. Gov't] *Nature*, *479*, 241–244.

Gell, D., & Jackson, S. P. (1999). Mapping of protein-protein interactions within the DNA-dependent protein kinase complex. *Nucleic Acids Research*, *27*, 3494–3502.

Goodarzi, A. A., Kurka, T., & Jeggo, P. A. (2011). KAP-1 phosphorylation regulates CHD3 nucleosome remodeling during the DNA double-strand break response. *Nature Structural & Molecular Biology*, *18*, 831–839.

Goodarzi, A. A., Noon, A. T., Deckbar, D., Ziv, Y., Shiloh, Y., Lobrich, M., et al. (2008). ATM signaling facilitates repair of DNA double-strand breaks associated with heterochromatin. *Molecular Cell*, *31*, 167–177.

Gospodinov, A., & Herceg, Z. (2012). Shaping chromatin for repair. *Mutation Research*, *752*, 45–60.

Gozalbo-Lopez, B., Andrade, P., Terrados, G., de Andres, B., Serrano, N., Cortegano, I., et al. (2009). A role for DNA polymerase mu in the emerging DJH rearrangements of the postgastrulation mouse embryo. *Molecular and Cellular Biology*, *29*, 1266–1275.

Grawunder, U., Wilm, M., Wu, X., Kulesza, P., Wilson, T. E., Mann, M., et al. (1997). Activity of DNA ligase IV stimulated by complex formation with XRCC4 protein in mammalian cells. *Nature*, *388*, 492–495.

Grundy, G. J., Rulten, S. L., Zeng, Z., Arribas-Bosacoma, R., Iles, N., Manley, K., et al. (2012). APLF promotes the assembly and activity of non-homologous end joining protein complexes. *The EMBO Journal, 32*, 112–125.

Hammel, M., Rey, M., Yu, Y., Mani, R. S., Classen, S., Liu, M., et al. (2011). XRCC4 protein interactions with XRCC4-like factor (XLF) create an extended grooved scaffold for DNA ligation and double strand break repair. *The Journal of Biological Chemistry, 286*, 32638–32650.

Hammel, M., Yu, Y., Mahaney, B. L., Cai, B., Ye, R., Phipps, B. M., et al. (2010). Ku and DNA-dependent protein kinase dynamic conformations and assembly regulate DNA binding and the initial non-homologous end joining complex. *The Journal of Biological Chemistry, 285*, 1414–1423.

Harreman, M., Taschner, M., Sigurdsson, S., Anindya, R., Reid, J., Somesh, B., et al. (2009). Distinct ubiquitin ligases act sequentially for RNA polymerase II polyubiquitylation. *Proceedings of the National Academy of Sciences of the United States of America, 106*, 20705–20710.

Hartley, K. O., Gell, D., Smith, G. C. M., Zhang, H., Divecha, N., Connelly, M. A., et al. (1995). DNA-dependent protein kinase catalytic subunit: A relative of phosphatidylinositol 3-kinase and the ataxia telangiectasia gene product. *Cell, 82*, 849–856.

Hilario, J., & Kowalczykowski, S. C. (2010). Visualizing protein-DNA interactions at the single-molecule level. *Current Opinion in Chemical Biology, 14*, 15–22.

Hiom, K., & Gellert, M. (1997). A stable RAG1-RAG2-DNA complex that is active in V(D)J cleavage. *Cell, 88*, 65–72.

Holthausen, J. T., van Loenhout, M. T., Sanchez, H., Ristic, D., van Rossum-Fikkert, S. E., Modesti, M., et al. (2011). Effect of the BRCA2 CTRD domain on RAD51 filaments analyzed by an ensemble of single molecule techniques. *Nucleic Acids Research, 39*, 6558–6567.

Horsfield, J. A., Print, C. G., & Monnich, M. (2012). Diverse developmental disorders from the one ring: Distinct molecular pathways underlie the cohesinopathies. *Frontiers in Genetics, 3*, 171.

Hu, Y., Scully, R., Sobhian, B., Xie, A., Shestakova, E., & Livingston, D. M. (2011). RAP80-directed tuning of BRCA1 homologous recombination function at ionizing radiation-induced nuclear foci. *Genes & Development, 25*, 685–700.

Huyen, Y., Zgheib, O., Ditullio, R. A., Jr., Gorgoulis, V. G., Zacharatos, P., Petty, T. J., et al. (2004). Methylated lysine 79 of histone H3 targets 53BP1 to DNA double-strand breaks. *Nature, 432*, 406–411.

Iacovoni, J. S., Caron, P., Lassadi, I., Nicolas, E., Massip, L., Trouche, D., et al. (2010). High-resolution profiling of gammaH2AX around DNA double strand breaks in the mammalian genome. *The EMBO Journal, 29*, 1446–1457.

Iliakis, G. (2009). Backup pathways of NHEJ in cells of higher eukaryotes: Cell cycle dependence. *Radiotherapy and Oncology, 92*, 310–315.

Iliakis, G., Metzger, L., Denko, N., & Stamato, T. D. (1991). Detection of DNA double strand breaks in synchronous cultures of CHO cells by means of asymmetric field inversion gel electrophoresis. *International Journal of Radiation Biology, 59*, 321–341.

Ip, S. C., Rass, U., Blanco, M. G., Flynn, H. R., Skehel, J. M., & West, S. C. (2008). Identification of Holliday junction resolvases from humans and yeast. *Nature, 456*, 357–361.

Ivanov, A. V., Peng, H., Yurchenko, V., Yap, K. L., Negorev, D. G., Schultz, D. C., et al. (2007). PHD domain-mediated E3 ligase activity directs intramolecular sumoylation of an adjacent bromodomain required for gene silencing. *Molecular Cell, 28*, 823–837.

Jackson, S. P., & Bartek, J. (2009). The DNA-damage response in human biology and disease. *Nature, 461*, 1071–1078.

Jakob, B., Splinter, J., Conrad, S., Voss, K. O., Zink, D., Durante, M., et al. (2011). DNA double-strand breaks in heterochromatin elicit fast repair protein recruitment, histone

H2AX phosphorylation and relocation to euchromatin. *Nucleic Acids Research, 39,* 6489–6499.

Jeggo, P., & Lavin, M. F. (2009). Cellular radiosensitivity: How much better do we understand it? *International Journal of Radiation Biology, 85,* 1061–1081.

Jensen, R. B., Carreira, A., & Kowalczykowski, S. C. (2010). Purified human BRCA2 stimulates RAD51-mediated recombination. *Nature, 467,* 678–683.

Johnson, R. D., & Jasin, M. (2000). Sister chromatid gene conversion is a prominent double-strand break repair pathway in mammalian cells. *The EMBO Journal, 19,* 3398–3407.

Kass, E. M., & Jasin, M. (2010). Collaboration and competition between DNA double-strand break repair pathways. *FEBS Letters, 584,* 3703–3708.

Kastan, M. B., & Bartek, J. (2004). Cell-cycle checkpoints and cancer. *Nature, 432,* 316–323.

Kastan, M. B., Zhan, Q., El-Deiry, W. S., Carrier, F., Jacks, T., Walsh, W. V., et al. (1992). A mammalian cell cycle checkpoint pathway utilizing p53 and GADD45 is defective in ataxia-telangiectasia. *Cell, 71,* 587–597.

Kim, H., Chen, J., & Yu, X. (2007). Ubiquitin-binding protein RAP80 mediates BRCA1-dependent DNA damage response. *Science, 316,* 1202–1205.

Kim, J., Daniel, J., Espejo, A., Lake, A., Krishna, M., Xia, L., et al. (2006). Tudor, MBT and chromo domains gauge the degree of lysine methylation. *EMBO Reports, 7,* 397–403.

Kim, J. A., Kruhlak, M., Dotiwala, F., Nussenzweig, A., & Haber, J. E. (2007). Heterochromatin is refractory to gamma-H2AX modification in yeast and mammals. *The Journal of Cell Biology, 178,* 209–218.

Koch, C. A., Agyei, R., Galicia, S., Metalnikov, P., O'Donnell, P., Starostine, A., et al. (2004). Xrcc4 physically links DNA end processing by polynucleotide kinase to DNA ligation by DNA ligase IV. *The EMBO Journal, 23,* 3874–3885.

Kruhlak, M., Crouch, E. E., Orlov, M., Montano, C., Gorski, S. A., Nussenzweig, A., et al. (2007). The ATM repair pathway inhibits RNA polymerase I transcription in response to chromosome breaks. *Nature, 447,* 730–734.

Lan, L., Ui, A., Nakajima, S., Hatakeyama, K., Hoshi, M., Watanabe, R., et al. (2010). The ACF1 complex is required for DNA double-strand break repair in human cells. *Molecular Cell, 40,* 976–987.

Larsen, N. B., & Hickson, I. D. (2013). RecQ helicases: Conserved guardians of genomic integrity. *Advances in Experimental Medicine and Biology, 767,* 161–184.

Larsen, D. H., Poinsignon, C., Gudjonsson, T., Dinant, C., Payne, M. R., Hari, F. J., et al. (2010). The chromatin-remodeling factor CHD4 coordinates signaling and repair after DNA damage. *The Journal of Cell Biology, 190,* 731–740.

Lee, J. H., Goodarzi, A. A., Jeggo, P. A., & Paull, T. T. (2010). 53BP1 promotes ATM activity through direct interactions with the MRN complex. *The EMBO Journal, 29,* 574–585.

Lee, J., Thompson, J. R., Botuyan, M. V., & Mer, G. (2008). Distinct binding modes specify the recognition of methylated histones H3K4 and H4K20 by JMJD2A-tudor. *Nature Structural & Molecular Biology, 15,* 109–111.

Lewis, S. M. (1994). P nucleotide insertions and the resolution of hairpin DNA structures in mammalian cells. *Proceedings of the National Academy of Sciences of the United States of America, 91,* 1332–1336.

Li, Y., Chirgadze, D. Y., Bolanos-Garcia, V. M., Sibanda, B. L., Davies, O. R., Ahnesorg, P., et al. (2008). Crystal structure of human XLF/Cernunnos reveals unexpected differences from XRCC4 with implications for NHEJ. *The EMBO Journal, 27,* 290–300.

Lieber, M. R. (2010). The mechanism of double-strand DNA break repair by the non-homologous DNA end-joining pathway. *Annual Review of Biochemistry, 79,* 181–211.

Liu, J., & Baynam, G. (2010). Cornelia de Lange syndrome. [Review] *Advances in Experimental Medicine and Biology, 685,* 111–123.

Lobrich, M., & Jeggo, P. A. (2007). The impact of a negligent G2/M checkpoint on genomic instability and cancer induction. *Nature Reviews. Cancer, 7,* 861–869.

Lucas, D., Escudero, B., Ligos, J. M., Segovia, J. C., Estrada, J. C., Terrados, G., et al. (2009). Altered hematopoiesis in mice lacking DNA polymerase mu is due to inefficient double-strand break repair. *PLoS Genetics, 5,* e1000389.

Luijsterburg, M. S., Acs, K., Ackermann, L., Wiegant, W. W., Bekker-Jensen, S., Larsen, D. H., et al. (2012). A new non-catalytic role for ubiquitin ligase RNF8 in unfolding higher-order chromatin structure. *The EMBO Journal, 31,* 2511–2527.

Luijsterburg, M. S., & van Attikum, H. (2011). Chromatin and the DNA damage response: The cancer connection. *Molecular Oncology, 5,* 349–367.

Ma, Y., Pannicke, U., Schwarz, K., & Lieber, M. R. (2002). Hairpin opening and overhang processing by an Artemis/DNA-dependent protein kinase complex in nonhomologous end joining and V(D)J recombination. *Cell, 108,* 781–794.

Malik, S., Bagla, S., Chaurasia, P., Duan, Z., & Bhaumik, S. R. (2008). Elongating RNA polymerase II is disassembled through specific degradation of its largest but not other sub-units in response to DNA damage in vivo. *The Journal of Biological Chemistry, 283,* 6897–6905.

Mallette, F. A., Mattiroli, F., Cui, G., Young, L. C., Hendzel, M. J., Mer, G., et al. (2012). RNF8- and RNF168-dependent degradation of KDM4A/JMJD2A triggers 53BP1 recruitment to DNA damage sites. *The EMBO Journal, 31,* 1865–1878.

Mani, R. S., Fanta, M., Karimi-Busheri, F., Silver, E., Virgen, C. A., Caldecott, K. W., et al. (2007). XRCC1 stimulates polynucleotide kinase by enhancing its damage discrimination and displacement from DNA repair intermediates. *The Journal of Biological Chemistry, 282,* 28004–28013.

Manis, J. P., Morales, J. C., Xia, Z., Kutok, J. L., Alt, F. W., & Carpenter, P. B. (2004). 53BP1 links DNA damage-response pathways to immunoglobulin heavy chain class-switch recombination. *Nature Immunology, 5,* 481–487.

Mari, P. O., Florea, B. I., Persengiev, S. P., Verkaik, N. S., Bruggenwirth, H. T., Modesti, M., et al. (2006). Dynamic assembly of end-joining complexes requires interaction between Ku70/80 and XRCC4. *Proceedings of the National Academy of Sciences of the United States of America, 103,* 18597–18602.

Marples, B., & Collis, S. J. (2008). Low-dose hyper-radiosensitivity: Past, present, and future. *International Journal of Radiation Oncology, Biology, and Physics, 70,* 1310–1318.

Matos, J., Blanco, M. G., Maslen, S., Skehel, J. M., & West, S. C. (2011). Regulatory control of the resolution of DNA recombination intermediates during meiosis and mitosis. [Research Support, Non-U.S. Gov't]. *Cell, 147,* 158–172.

Mazin, A. V., Mazina, O. M., Bugreev, D. V., & Rossi, M. J. (2010). Rad54, the motor of homologous recombination. *DNA Repair, 9,* 286–302.

Mazon, G., Mimitou, E. P., & Symington, L. S. (2010). SnapShot: Homologous recombination in DNA double-strand break repair. *Cell, 142,* 646, 646 e641.

Meek, K., Douglas, P., Cui, X., Ding, Q., & Lees-Miller, S. P. (2007). trans Autophosphorylation at DNA-dependent protein kinase's two major autophosphorylation site clusters facilitates end processing but not end joining. *Molecular and Cellular Biology, 27,* 3881–3890.

Meerang, M., Ritz, D., Paliwal, S., Garajova, Z., Bosshard, M., Mailand, N., et al. (2011). The ubiquitin-selective segregase VCP/p97 orchestrates the response to DNA double-strand breaks. *Nature Cell Biology, 13,* 1376–1382.

Messick, T. E., & Greenberg, R. A. (2009). The ubiquitin landscape at DNA double-strand breaks. *The Journal of Cell Biology, 187,* 319–326.

Mladenov, E., & Iliakis, G. (2011). Induction and repair of DNA double strand breaks: The increasing spectrum of non-homologous end joining pathways. *Mutation Research, 711,* 61–72.

Moyal, L., Lerenthal, Y., Gana-Weisz, M., Mass, G., So, S., Wang, S. Y., et al. (2011). Requirement of ATM-dependent monoubiquitylation of histone H2B for timely repair of DNA double-strand breaks. *Molecular Cell, 41*, 529–542.

Moynahan, M. E., & Jasin, M. (2010). Mitotic homologous recombination maintains genomic stability and suppresses tumorigenesis. *Nature Reviews. Molecular Cell Biology, 11*, 196–207.

Muramatsu, M., Kinoshita, K., Fagarasan, S., Yamada, S., Shinkai, Y., & Honjo, T. (2000). Class switch recombination and hypermutation require activation-induced cytidine deaminase (AID), a potential RNA editing enzyme. *Cell, 102*, 553–563.

Murr, R., Loizou, J. I., Yang, Y. G., Cuenin, C., Li, H., Wang, Z. Q., et al. (2006). Histone acetylation by Trrap-Tip60 modulates loading of repair proteins and repair of DNA double-strand breaks. [Research Support, Non-U.S. Gov't] *Nature Cell Biology, 8*, 91–99.

Nakamura, K., Kato, A., Kobayashi, J., Yanagihara, H., Sakamoto, S., Oliveira, D. V., et al. (2011). Regulation of homologous recombination by RNF20-dependent H2B ubiquitination. [Research Support, Non-U.S. Gov't] *Molecular Cell, 41*, 515–528.

Neal, J. A., & Meek, K. (2011). Choosing the right path: Does DNA-PK help make the decision? *Mutation Research, 711*, 73–86.

Noon, A. T., & Goodarzi, A. A. (2011). 53BP1-mediated DNA double strand break repair: Insert bad pun here. [Review] *DNA Repair, 10*, 1071–1076.

Noon, A. T., Shibata, A., Rief, N., Lobrich, M., Stewart, G. S., Jeggo, P. A., et al. (2010). 53BP1-dependent robust localized KAP-1 phosphorylation is essential for heterochromatic DNA double-strand break repair. *Nature Cell Biology, 12*, 177–184.

Nussenzweig, A., & Nussenzweig, M. C. (2007). A backup DNA repair pathway moves to the forefront. *Cell, 131*, 223–225.

O'Driscoll, M., Cerosaletti, K. M., Girard, P.-M., Dai, Y., Stumm, M., Kysela, B., et al. (2001). DNA Ligase IV mutations identified in patients exhibiting development delay and immunodeficiency. *Molecular Cell, 8*, 1175–1185.

O'Driscoll, M., & Jeggo, P. A. (2006). The role of double-strand break repair: Insights from human genetics. *Nature Reviews. Genetics, 7*, 45–54.

O'Driscoll, M., & Jeggo, P. A. (2008). CsA can induce DNA double-strand breaks: Implications for BMT regimens particularly for individuals with defective DNA repair. *Bone Marrow Transplantation, 41*, 983–989.

Panier, S., & Durocher, D. (2009). Regulatory ubiquitylation in response to DNA double-strand breaks. *DNA Repair (Amst), 8*, 436–443.

Pankotai, T., Bonhomme, C., Chen, D., & Soutoglou, E. (2012). DNAPKcs-dependent arrest of RNA polymerase II transcription in the presence of DNA breaks. *Nature Structural & Molecular Biology, 19*, 276–282.

Patterson-Fortin, J., Shao, G., Bretscher, H., Messick, T. E., & Greenberg, R. A. (2010). Differential regulation of JAMM domain deubiquitinating enzyme activity within the RAP80 complex. *The Journal of Biological Chemistry, 285*, 30971–30981.

Pei, H., Zhang, L., Luo, K., Qin, Y., Chesi, M., Fei, F., et al. (2011). MMSET regulates histone H4K20 methylation and 53BP1 accumulation at DNA damage sites. *Nature, 470*, 124–128.

Petermann, E., & Helleday, T. (2010). Pathways of mammalian replication fork restart. *Nature Reviews. Molecular Cell Biology, 11*, 683–687.

Polo, S. E., & Jackson, S. P. (2011). Dynamics of DNA damage response proteins at DNA breaks: A focus on protein modifications. *Genes & Development, 25*, 409–433.

Polo, S. E., Kaidi, A., Baskcomb, L., Galanty, Y., & Jackson, S. P. (2010). Regulation of DNA-damage responses and cell-cycle progression by the chromatin remodelling factor CHD4. *The EMBO Journal, 29*, 3130–3139.

Rassool, F. V., & Tomkinson, A. E. (2010). Targeting abnormal DNA double strand break repair in cancer. *Cellular and Molecular Life Sciences, 67*, 3699–3710.

Riballo, E., Kuhne, M., Rief, N., Doherty, A., Smith, G. C., Recio, M. J., et al. (2004). A pathway of double-strand break rejoining dependent upon ATM, Artemis, and proteins locating to gamma-H2AX foci. *Molecular Cell, 16*, 715–724.

Riballo, E., Woodbine, L., Stiff, T., Walker, S. A., Goodarzi, A. A., & Jeggo, P. A. (2009). XLF-Cernunnos promotes DNA ligase IV-XRCC4 re-adenylation following ligation. *Nucleic Acids Research, 37*, 482–492.

Ropars, V., Drevet, P., Legrand, P., Baconnais, S., Amram, J., Faure, G., et al. (2011). Structural characterization of filaments formed by human Xrcc4-Cernunnos/XLF complex involved in nonhomologous DNA end-joining. *Proceedings of the National Academy of Sciences of the United States of America, 108*, 12663–12668.

Rulten, S. L., Fisher, A. E. O., Robert, I., Zuma, M. C., Rouleau, M., Ju, L., et al. (2011). PARP-3 and APLF function together to accelerate nonhomologous end-joining. *Molecular Cell, 41*, 33–45.

Schultz, D. C., Friedman, J. R., & Rauscher, F. J., 3rd. (2001). Targeting histone deacetylase complexes via KRAB-zinc finger proteins: The PHD and bromodomains of KAP-1 form a cooperative unit that recruits a novel isoform of the Mi-2alpha subunit of NuRD. *Genes & Development, 15*, 428–443.

Segal-Raz, H., Mass, G., Baranes-Bachar, K., Lerenthal, Y., Wang, S. Y., Chung, Y. M., et al. (2011). ATM-mediated phosphorylation of polynucleotide kinase/phosphatase is required for effective DNA double-strand break repair. *EMBO Reports, 12*, 713–719.

Shanbhag, N. M., Rafalska-Metcalf, I. U., Balane-Bolivar, C., Janicki, S. M., & Greenberg, R. A. (2010). ATM-dependent chromatin changes silence transcription in cis to DNA double-strand breaks. *Cell, 141*, 970–981.

Shao, G., Lilli, D. R., Patterson-Fortin, J., Coleman, K. A., Morrissey, D. E., & Greenberg, R. A. (2009). The Rap80-BRCC36 de-ubiquitinating enzyme complex antagonizes RNF8-Ubc13-dependent ubiquitination events at DNA double strand breaks. *Proceedings of the National Academy of Sciences of the United States of America, 106*, 3166–3171.

Shibata, A., Barton, O., Noon, A. T., Dahm, K., Deckbar, D., Goodarzi, A. A., et al. (2009). The maintainance of ATM dependant G2/M checkpoint arrest following exposure to ionizing radiation. *Acta Medica Nagasakiensia*, 19–21.

Shibata, A., Barton, O., Noon, A. T., Dahm, K., Deckbar, D., Goodarzi, A. A., et al. (2010). Role of ATM and the damage response mediator proteins 53BP1 and MDC1 in the maintenance of G(2)/M checkpoint arrest. *Molecular and Cellular Biology, 30*, 3371–3383.

Shibata, A., Conrad, S., Birraux, J., Geuting, V., Barton, O., Ismail, A., et al. (2011). Factors determining DNA double-strand break repair pathway choice in G2 phase. *The EMBO Journal, 30*, 1079–1092.

Sibanda, B. L., Chirgadze, D. Y., & Blundell, T. L. (2010). Crystal structure of DNA-PKcs reveals a large open-ring cradle comprised of HEAT repeats. *Nature, 463*, 118–121.

Sibanda, B. L., Critchlow, S. E., Begun, J., Pei, X. Y., Jackson, S. P., Blundell, T. L., et al. (2001). Crystal structure of an Xrcc4-DNA ligase IV complex. *Natural Structural Biology, 8*, 1015–1019.

Simsek, D., Brunet, E., Wong, S. Y., Katyal, S., Gao, Y., McKinnon, P. J., et al. (2011). DNA ligase III promotes alternative nonhomologous end-joining during chromosomal translocation formation. *PLoS Genetics, 7*, e1002080.

Simsek, D., & Jasin, M. (2010). Alternative end-joining is suppressed by the canonical NHEJ component Xrcc4-ligase IV during chromosomal translocation formation. *Nature Structural & Molecular Biology, 17*, 410–416.

Singleton, B. K., Torres-Arzayus, M. I., Rottinghaus, S. T., Taccioli, G. E., & Jeggo, P. A. (1999). The C terminus of Ku80 activates the DNA-dependent protein kinase catalytic subunit. *Molecular and Cellular Biology, 19*, 3267–3277.

Sobhian, B., Shao, G., Lilli, D. R., Culhane, A. C., Moreau, L. A., Xia, B., et al. (2007). RAP80 targets BRCA1 to specific ubiquitin structures at DNA damage sites. *Science, 316*, 1198–1202.

Soubeyrand, S., Pope, L., Pakuts, B., & Hache, R. J. (2003). Threonines 2638/2647 in DNA-PK are essential for cellular resistance to ionizing radiation. *Cancer Research, 63*, 1198–1201.

Staszewski, O., Baker, R. E., Ucher, A. J., Martier, R., Stavnezer, J., & Guikema, J. E. (2011). Activation-induced cytidine deaminase induces reproducible DNA breaks at many non-Ig Loci in activated B cells. *Molecular Cell, 41*, 232–242.

Stavnezer, J., Guikema, J. E., & Schrader, C. E. (2008). Mechanism and regulation of class switch recombination. *Annual Review of Immunology, 26*, 261–292.

Stewart, G. S., Maser, R. S., Stankovic, T., Bressan, D. A., Kaplan, M. I., Jaspers, N. G., et al. (1999). The DNA double-strand break repair gene hMRE11 is mutated in individuals with an ataxia-telangiectasia-like disorder. *Cell, 99*, 577–587.

Stewart, G. S., Panier, S., Townsend, K., Al-Hakim, A. K., Kolas, N. K., Miller, E. S., et al. (2009). The RIDDLE syndrome protein mediates a ubiquitin-dependent signaling cascade at sites of DNA damage. *Cell, 136*, 420–434.

Stiff, T., O'Driscoll, M., Rief, N., Iwabuchi, K., Lobrich, M., & Jeggo, P. A. (2004). ATM and DNA-PK function redundantly to phosphorylate H2AX following exposure to ioninsing radiation. *Cancer Research, 64*, 2390–2396.

Sun, Y., Jiang, X., Chen, S., Fernandes, N., & Price, B. D. (2005). A role for the Tip60 histone acetyltransferase in the acetylation and activation of ATM. *Proceedings of the National Academy of Sciences of the United States of America, 102*, 13182–13187.

Sun, Y., Jiang, X., Xu, Y., Ayrapetov, M. K., Moreau, L. A., Whetstine, J. R., et al. (2009). Histone H3 methylation links DNA damage detection to activation of the tumour suppressor Tip60. *Nature Cell Biology, 11*(11), 1376–1382.

Sun, Y., Xu, Y., Roy, K., & Price, B. D. (2007). DNA damage-induced acetylation of lysine 3016 of ATM activates ATM kinase activity. *Molecular and Cellular Biology, 27*, 8502–8509.

Svendsen, J. M., & Harper, J. W. (2010). GEN1/Yen1 and the SLX4 complex: Solutions to the problem of Holliday junction resolution. *Genes & Development, 24*, 521–536.

Symington, L. S., & Gautier, J. (2011). Double-strand break end resection and repair pathway choice. *Annual Review of Genetics, 45*, 247–271.

Teo, S.-H., & Jackson, S. P. (1997). Identification of *Saccharomyces cerevisiae* DNA ligase IV: Involvement in DNA double-strand break repair. *The EMBO Journal, 16*, 4788–4795.

Thompson, L. H. (2012). Recognition, signaling, and repair of DNA double-strand breaks produced by ionizing radiation in mammalian cells: The molecular choreography. *Mutation Research, 751*, 158–246.

Urquhart, A. J., Gatei, M., Richard, D. J., & Khanna, K. K. (2011). ATM mediated phosphorylation of CHD4 contributes to genome maintenance. *Genome Integrity, 2*, 1.

Wahl, G. M., Linke, S. P., Paulson, T. G., & Huang, L. C. (1997). Maintaining genetic stability through TP53 mediated checkpoint control. *Cancer Surveys, 29*, 183–219.

Walker, J. R., Corpina, R. A., & Goldberg, J. (2001). Structure of the Ku heterodimer bound to DNA and its implications for double-strand break repair. *Nature, 412*, 607–614.

Waltes, R., Kalb, R., Gatei, M., Kijas, A. W., Stumm, M., Sobeck, A., et al. (2009). Human RAD50 deficiency in a Nijmegen breakage syndrome-like disorder. [Case Reports] *The American Journal of Human Genetics, 84*, 605–616.

Wang, B., & Elledge, S. J. (2007). Ubc13/Rnf8 ubiquitin ligases control foci formation of the Rap80/Abraxas/Brca1/Brcc36 complex in response to DNA damage. *Proceedings of the National Academy of Sciences of the United States of America, 104,* 20759–20763.

Wang, H., Perrault, A. R., Takeda, Y., Qin, W., & Iliakis, G. (2003). Biochemical evidence for Ku-independent backup pathways of NHEJ. *Nucleic Acids Research, 31,* 5377–5388.

Wang, H., Rosidi, B., Perrault, R., Wang, M., Zhang, L., Windhofer, F., et al. (2005). DNA ligase III as a candidate component of backup pathways of nonhomologous end joining. *Cancer Research, 65,* 4020–4030.

Wang, M., Wu, W., Rosidi, B., Zhang, L., Wang, H., & Iliakis, G. (2006). PARP-1 and Ku compete for repair of DNA double strand breaks by distinct NHEJ pathways. *Nucleic Acids Research, 34,* 6170–6182.

Ward, I. M., Reina-San-Martin, B., Olaru, A., Minn, K., Tamada, K., Lau, J. S., et al. (2004). 53BP1 is required for class switch recombination. *The Journal of Cell Biology, 165,* 459–464.

Wechsler, T., Newman, S., & West, S. C. (2011). Aberrant chromosome morphology in human cells defective for Holliday junction resolution. *Nature, 471,* 642–646.

Wilson, T. E., Grawunder, U., & Lieber, M. R. (1997). Yeast DNA ligase IV mediates nonhomologous DNA end joining. *Nature, 388,* 495–498.

Wu, L., & Hickson, I. D. (2003). The Bloom's syndrome helicase suppresses crossing over during homologous recombination. *Nature, 426,* 870–874.

Xie, A., Hartlerode, A., Stucki, M., Odate, S., Puget, N., Kwok, A., et al. (2007). Distinct roles of chromatin-associated proteins MDC1 and 53BP1 in mammalian double-strand break repair. *Molecular Cell, 28,* 1045–1057.

Xu, Y., Sun, Y., Jiang, X., Ayrapetov, M. K., Moskwa, P., Yang, S., et al. (2010). The p400 ATPase regulates nucleosome stability and chromatin ubiquitination during DNA repair. *The Journal of Cell Biology, 191,* 31–43.

Yan, C. T., Boboila, C., Souza, E. K., Franco, S., Hickernell, T. R., Murphy, M., et al. (2007). IgH class switching and translocations use a robust non-classical end-joining pathway. *Nature, 449,* 478–482.

Yun, M. H., & Hiom, K. (2009). CtIP-BRCA1 modulates the choice of DNA double-strand-break repair pathway throughout the cell cycle. *Nature, 459,* 460–463.

Ziv, Y., Bielopolski, D., Galanty, Y., Lukas, C., Taya, Y., Schultz, D. C., et al. (2006). Chromatin relaxation in response to DNA double-strand breaks is modulated by a novel ATM- and KAP-1 dependent pathway. *Nature Cell Biology, 8,* 870–876.

Zolner, A. E., Abdou, I., Ye, R., Mani, R. S., Fanta, M., Yu, Y., et al. (2011). Phosphorylation of polynucleotide kinase/ phosphatase by DNA-dependent protein kinase and ataxia-telangiectasia mutated regulates its association with sites of DNA damage. *Nucleic Acids Research, 39,* 9224–9237.

Zucca, E., Bertoletti, F., Wimmer, U., Ferrari, E., Mazzini, G., Khoronenkova, S., et al. (2013). Silencing of human DNA polymerase lambda causes replication stress and is synthetically lethal with an impaired S phase checkpoint. *Nucleic Acids Research, 14,* 229–241.

CHAPTER TWO

Biological Activity and Biotechnological Aspects of Locked Nucleic Acids

Karin E. Lundin[*,1], Torben Højland[†], Bo R. Hansen[‡], Robert Persson[‡],
Jesper B. Bramsen[§], Jørgen Kjems[§], Troels Koch[‡], Jesper Wengel[†],
C.I. Edvard Smith[*,1]

[*]Clinical Research Center, Department of Laboratory Medicine, Karolinska Institutet, Novum, Huddinge, Stockholm, Sweden
[†]NanoCAN, Department of Physics, Chemistry and Pharmacy, University of Southern Denmark, Odense, Denmark
[‡]Santaris Pharma A/S, Hørsholm, Denmark
[§]Interdisciplinary Nanoscience Center (iNANO), Department of Molecular Biology and Genetics, Aarhus University, Aarhus, Denmark
[1]Corresponding authors: e-mail address: karin.lundin@ki.se; edvard.smith@ki.se

Contents

Advances in Genetics, Volume 82
ISSN 0065-2660
http://dx.doi.org/10.1016/B978-0-12-407676-1.00002-0

Abstract

Locked nucleic acid (LNA) is one of the most promising new nucleic acid analogues that has been produced under the past two decades. In this chapter, we have tried to cover many of the different areas, where this molecule has been used to improve the function of synthetic oligonucleotides (ONs). The use of LNA in antisense ONs, including gapmers, splice-switching ONs, and siLNA, as well as antigene ONs, is reviewed. Pharmacokinetics as well as pharmacodynamics of LNA ONs and a description of selected compounds in, or close to, clinical testing are described. In addition, new LNA modifications and the adaptation of enzymes for LNA incorporation are reviewed. Such enzymes may become important for the development of stabilized LNA-containing aptamers.

1. BRIEF INTRODUCTION

The development of novel synthetic nucleotide chemistries has produced hundreds of new compounds in laboratories all over the world. While this has resulted in an enormous amount of new and exciting information, rather few of these chemistries have become more widely spread. Locked nucleic acid (LNA) represents one of the synthetic nucleotide chemistries, which has become widely used. It was originally developed in the late 1990s (Obika et al., 1997; Singh, Nielsen, Koshkin, & Wengel, 1998) and is characterized by the inclusion of a methylene bridge connecting the $2'$-oxygen and the $4'$-carbon atoms in the furanose ring. This locking effect restricts the conformation, which forms the basis for the enhanced hybridization properties of this compound.

In this review, we have tried to cover many aspects of LNA, from the basic chemistry and structure to its use as antigene and antisense reagents, enabling both degradation and alteration of nucleic acids inside cells. We further describe how LNA has been used to treat disease both in animal experimental models and humans. There has been a remarkable development in this field over the past few years and some of the authors have hands–on experience from this work. Other authors have pioneered the construction of new and powerful tools in biotechnology, where there are many applications and recent exciting data from trying to engineer enzymes, which can incorporate LNA-bases into growing oligo- or polynucleotide chains.

2. CHEMISTRY AND STRUCTURE

2.1. LNA characteristics and structure

LNA oligonucleotides (ONs) contain one or more LNA ON monomer(s) (Figure 2.1), which is a ribonucleotide analogue where the $2'$-oxygen and the

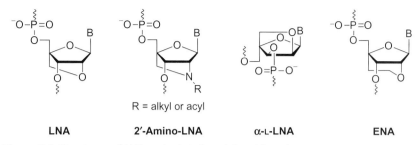

R = alkyl or acyl

LNA   **2′-Amino-LNA**   **α-L-LNA**   **ENA**

Figure 2.1 Structures of LNA and related nucleic acid analogues.

4′-carbon atoms are connected via a methylene bridge (Obika et al., 1997; Singh, Nielsen, et al., 1998). This bridge locks the sugar moiety in an N-type sugar ring conformation such that LNA can be considered an RNA mimic. This conformational restriction results in preorganization of the backbone of LNA ONs, which leads to energetically favorable duplex formation, in part via increased base stacking interactions (Kaur, Babu, & Maiti, 2007; Petersen et al., 2000). Generally, the melting temperature (T_m) of duplexes is raised by 2–8 °C per LNA nucleotide (nt) incorporated when compared to the corresponding unmodified duplexes, depending on the sequence context and number of modifications (Veedu & Wengel, 2010). This makes LNA the prime nt modification candidate for applications where high affinity is desirable. Importantly, custom ONs, with incorporations of LNA monomers are commercially available, also in combination with a wide variety of other nt modifications and labels. The phosphoramidite building blocks of LNA are likewise commercially available and can be used on an automated nucleic acid synthesizer.

The RNA-mimicking nature of LNA has been confirmed in NMR studies of duplexes containing LNA nts (Nielsen et al., 2004; Petersen, Bondensgaard, Wengel, & Jacobsen, 2002). When one or three LNA monomers were introduced into DNA ONs, the solution structures of duplexes with complementary RNA showed that the LNA monomers conformationally steer the sugar moieties of 3′-flanking nts toward an N-type conformation. In fact, three LNA monomers in a 9-mer DNA ON lead to an almost canonical A-type duplex geometry (Nielsen et al., 2004). A similar duplex conformation is obtained when fully modified LNA ONs binds to complementary RNA, whereas fully modified LNA ONs bind to complementary DNA to form a duplex that structurally resembles an RNA:DNA duplex (Nielsen et al., 2004).

In order to use ONs as drugs, they must be chemically modified since nucleases and renal filtration quickly clear the blood stream of unmodified

ONs. Unmodified small interfering RNA (siRNA) is almost completely degraded in less than a minute, while insertion of LNA nts has shown to increase serum stability of siRNAs considerably (Gao et al., 2009) as well as they can provide nucleolytic stability to single-stranded DNA-based antisense ONs (ASOs). In one example, positioning of three LNA monomers in both ends improved half-life in human serum more than 10 times up to 17 h (Kurreck, Wyszko, Gillen, & Erdmann, 2002). A fully modified LNA is not degraded at all by S1 endonuclease after 2 h (Frieden et al., 2003), but when a gap of DNA nts is introduced in the center, the ON becomes increasingly susceptible as the gap length increases. Two LNA monomers toward the 3′-end induce some protection of the ON from degradation by the 3′-exonuclease snake venom phosphodiesterase, whereas a singly LNA-modified ON is degraded as fast as the unmodified control (Morita et al., 2003). Adding LNA nts centrally in an ON does not seem to slow down degradation by snake venom phosphodiesterase (Nagahama, Veedu, & Wengel, 2009). Double-stranded decoy ONs are also protected by LNA (Crinelli, Bianchi, Gentilini, & Magnani, 2002) and endonuclease DNase I degradation is significantly impeded by introduction of one or two terminally placed LNA nts. In another study, degradation by exonuclease BAL-31 was slowed for almost all LNA ONs included in the study but the best stability in this case was observed with internal modification. Importantly, protection from nucleases may be achieved by employing phosphorothioate (PS) linkages, which are fully compatible with LNA modification. This is a strategy generally employed when using single-stranded LNA-containing ASOs *in vivo* (Gupta et al., 2010; Stein et al., 2010).

2.2. LNA analogues and conjugates

Many variations on the LNA skeleton have been made. LNA analogues have been reviewed elsewhere (Obika et al., 2010; Zhou & Chattopadhyaya, 2009) and only few will be mentioned here.

The oxygen in the oxymethylene bridge that locks the sugar moiety in the LNA nt monomer can be exchanged for other atoms. Particularly interesting is 2′-amino-LNA (Figure 2.1) in which the sugar moiety retains the favorable properties associated with the locked *N*-type furanose ring conformation such as high thermal stability against DNA and RNA complements (Singh, Kumar, & Wengel, 1998a). At the same time the amino group can be used as a chemical handle onto which conjugating groups can be attached, for example, to improve fluorescence, hybridization, or pharmacokinetic

(PK) properties. Indeed, a variety of functionalities have been attached to the 2′-amino-LNA skeleton. Amino acids that possess a net positive charge under physiological conditions were attached to improve the thermal stability of duplexes (Johannsen, Crispino, Wamberg, Kalra, & Wengel, 2011) and triplexes (Højland, Kumar, et al., 2007). Perylene was attached to generate a probe for *in vivo* detection of mRNA using an excitation wavelength that lowers the problem of cell autofluorescence (Astakhova, Korshun, Jahn, Kjems, & Wengel, 2008). Additional nucleobases were attached to produce double-headed monomers, but then no indication of Watson–Crick (WC) base pairing with the second nucleobase was observed (Umemoto, Wengel, & Madsen, 2009). Cholesterol was attached to improve the activity of an miRNA knockdown probe (Bryld & Lomholt, 2007), and pyrene for the detection of complementary nucleic acids (Hrdlicka, Babu, Sørensen, Harrit, & Wengel, 2005) or hybridization events (Hrdlicka, Babu, Sørensen, & Wengel, 2004), and anthracene to effect strand cross-linking by photoligation (Pasternak, Pasternak, Gupta, Veedu, & Wengel, 2011).

Another interesting analogue is α-ʟ-LNA, which is a diastereomer of LNA (Figure 2.1). Unlike LNA, α-ʟ-LNA structurally is a DNA mimic (Nielsen, Stein, & Petersen, 2003). However, like LNA, α-ʟ-LNA provides increases in thermal stability of duplexes (Sørensen et al., 2002) and triplexes (Kumar, Nielsen, Maiti, & Petersen, 2006) and protects ONs from degradation by nucleases to an even greater extent than LNA (Frieden et al., 2003). Another advantage of α-ʟ-LNA in comparison with LNA is its better compatibility to RNase H activation, which leads to cleavage of the RNA strand of complexes formed between ASOs and their RNA targets (Frieden et al., 2003). Unfortunately, α-ʟ-LNA ONs have only to very limited extend been explored for therapeutic or biotechnological applications.

Ethylene-bridged nucleic acid (ENA, Figure 2.1) is a homologue of LNA, in which the bridge between the 2′- and 4′-carbon atoms has been extended by a methylene group (Morita et al., 2002). Like LNA, the sugar moiety in ENA is locked in an *N*-type conformation, but ENA nt monomers generally do not provide quite the same level of thermal stability to duplexes as LNA, although triplexes are stabilized significantly (Koizumi et al., 2003). As for many other LNA analogues, 3′-end modification of an ON with an ENA monomer induces better protection against snake venom phosphodiesterase than similar modification with an LNA monomer (Morita et al., 2003). However, as biological applications with LNA ONs generally are performed with the so-called gapmer- or mixmer-type oligomers in which several LNA nts are positioned in each end or LNA nts are dispersed equally throughout the sequence, respectively, nucleolytic stability is sufficient and often even further improved by inclusion of PS linkages.

3. BIOLOGICAL ACTIVITY OF LNA-CONTAINING ONs

Synthetic ONs with the capacity to base pair with natural nucleic acids in a highly sequence-specific manner possess excellent properties as gene regulatory agents, with very specific biological effects. Among the different chemistries, LNA-modified ONs have been extensively utilized in several areas: (i) classical antisense approaches to induce RNA degradation by RNase H, (ii) siRNA mediated degradation where the ONs are loaded into the RNA-induced silencing complex (RISC), (iii) altered pre-mRNA splicing, (iv) blocking of microRNA (anti-miR), and (v) antigene reagents to block transcription of a specific gene. In this chapter, we have tried to cover all of the mentioned areas except the anti-miR usage, which has been extensively reviewed very recently (Stenvang, Petri, Lindow, Obad, & Kauppinen, 2012; Thorsen, Obad, Jensen, Stenvang, & Kauppinen, 2012).

3.1. Antisense LNA

Depending on the target, and by which mechanism inhibition of RNA expression is most effective, LNA designs are used in two different constructs: mixmers and gapmers. In a mixmer, the LNA residues are dispersed throughout the sequence of the ON, while in a gapmer, two LNA nt segments at either end of the ON are separated by a central DNA segment, usually as PS. To inhibit mRNA expression, ON gapmer designs are the most potent. The reason is that the central DNA/PS segment, longer than 7–8 DNA nts, recruits the RNA cleaving enzyme RNase H when the ON is hybridized to the mRNA. For microRNA inhibition, the mixmer is the most effective design (Davis, Lollo, Freier, & Esau, 2006; Elmen, Lindow, Silahtaroglu, et al., 2008; Koch, Rosenbohm, Hansen, Straarup, & Kauppinen, 2008).

3.1.1 Affinity and potency

The affinity increase with every LNA nt substitution in ONs is perhaps the most important property of LNA. When LNA nts are phosphorothioated, the affinity increase appears to be even higher (up to 9–10 °C per LNA modification compared to PS alone) (Wengel et al., 1999). Studies have shown ΔT_m of mismatches in the range from $-21\,^{\circ}$C for T/C mismatches to ΔT_m of $-14\,^{\circ}$C for a T/G mismatch (Koshkin et al., 1998; McTigue, Peterson, & Kahn, 2004). However, the ΔT_m can vary from a few degree Celsius to more than $20\,^{\circ}$C depending on nt mismatch and sequence context (McTigue et al., 2004). By using LNAs, the required and optimal high binding affinity for any given ON to its complementary target RNA can be obtained by shorter molecules, and this is the reason for that LNAs have, from the very beginning, been designed shorter than older ASO generations (Frieden et al., 2003; Obika, 2004; Wahlestedt et al., 2000).

From a therapeutic point of view, the fact that LNA nts hybridize efficiently in combination with PS internucleoside linkages is of particular interest. It is well known that compared to the native phosphodiesters, PS provides improved PK and cellular uptake properties due to increased protein binding (Moschos et al., 2011; Reyes-Reyes, Teng, & Bates, 2010; Rockwell et al., 1997). PS ONs also remain longer in circulation after *in vivo* administration and associate better with cellular membranes. It has been demonstrated in many experiments that the affinity increase of LNA nucleotides, combined with increased nuclease resistance/protein binding of PS, is found for nearly all sequences (Koch & Ørum, 2007), a set of properties that is very important for making efficient ASOs (Koch & Ørum, 2007; Koch et al., 2008).

Binding affinity is an essential feature for any target/ligand interaction. We and others have shown that the high RNA binding affinity of LNA is linked to higher potency (Koch et al., 2008; Straarup et al., 2010; Yamamoto, Nakatani, Narukawa, & Obika, 2011). However, affinity above a certain compound-and mechanism-specific threshold (Straarup et al., 2010) may not add further to the activity if it compromises other potency-limiting factors. Examples of such factors are the rate of RNase H recruitment, tissue/cellular uptake, and/or tissue distribution. For instance, a fully LNA-modified 14-mer targeting the coding region of a transcript may not inhibit protein synthesis, since the ribosome simply removes the hybridized ON during translation despite the very high RNA affinity of the 14-mer (Braasch, Liu, & Corey, 2002; Obad et al., 2011). If some of the central LNA nucleosides (7–8 nts) are substituted with DNA nts, the 14-mer becomes an LNA gapmer. The hybridized ON will then recruit RNase H to the targeted mRNA, which will be cleaved and degraded (Braasch et al., 2002). The binding affinity of the 14-mer gapmer will be lower than the fully LNA-modified 14-mer, but the former, subsequent to RNase H-mediated mRNA cleavage, will exhibit potent downregulation of the target. In conclusion, high activity for a given compound is dependent on many factors relating to ON design, target class and mechanisms of action that all have to be balanced.

3.1.2 Cellular uptake and gymnosis

Transfection has for decades been the preferred procedure for intracellular delivery of ONs (Figure 2.2). The lipid bilayer membrane of the cell constitutes a solid barrier for the anionic ONs and the use of cationic lipids or conjugated ligands (e.g., cholesterol; Lebedeva, Benimetskaya, Stein, & Vilenchik, 2000) has been the only robust procedure to cross this barrier (Akhtar, Basu, Wickstrom, & Juliano, 1991; Alam et al., 2010; Lebedeva et al., 2000). However, it has also been demonstrated that ONs could be taken up by cells in the *absence* of transfection agents or conjugates (Alam et al., 2010; Manoharan, 2002), but it was also noted that the ASOs were confined to dense intracellular structures identified as endosomes/lysosomes (Figure 2.2). Uptake and internalization by this mechanism, the endocytic pathway, did

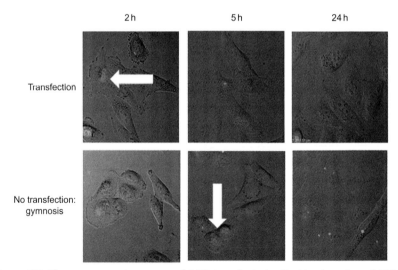

Figure 2.2 Fluorescence microscopy of ASO transfected cells. Live imaging of 518A2 cells after oligonucleotide treatment for 2, 5, and 24 h. In the upper panel, a fluorescently labeled LNA (10 nM) has been transfected into the cells to a spread nuclear distribution as indicated by white arrow. In the lower panel, the same LNA (2.5 µM) has been delivered without transfection agents via "gymnosis." The ONs are distributed in a speckled pattern, indicating uptake via vesicular compartments like the endocytic pathway. See arrow in lower panel. (For color version of this figure, the reader is referred to the online version of this chapter.)

not lead to any notable silencing activity for ASOs (Alam et al., 2010). The explanation for this was—and is—that the ONs may be "trapped" in the endosomes and not available for the silencing processes (Eckstein, 2007).

The uptake pattern for LNAs mimic what has previously been observed with other ONs. When nonconjugated LNAs were incubated without the addition of delivery vehicles (i.e., "naked"), they were internalized and distributed in a pattern that resembled uptake via the endocytic pathway (Stein et al., 2010; Figure 2.2). Over time the LNAs "traffic" through the cytoplasm and change from an initially highly speckled distribution to a more condensed presence. After approximately 10 days they are getting exocytosed and exported by the cells.

3.1.2.1 Uptake mechanisms

In the experiment illustrated in Figure 2.3, it was noted that the dynamics of LNAs taken up by the cells was very different when comparing transfection and unassisted uptake delivery conditions. When fluorescently labeled LNA

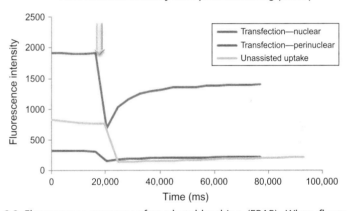

Fluorescence recovery after photobleaching (FRAP)

Figure 2.3 Fluorescence recovery after photobleaching (FRAP). When fluorescently labeled LNA (Flu-LNA) was transfected into cells the majority of the fluorescence accumulated in the nucleus. After laser photobleaching, partial restoration of the fluorescence was noted with time. When Flu-LNAs concentrated in perinuclear bodies was bleached, no restoration of fluorescence was observed. When Flu-LNA was taken up by gymnosis (unassisted uptake), the majority of the fluorescence was found in cytoplasmic-confined bodies. After photo bleaching of these loci, the fluorescence was not restored. Apparently intact fluorescently labeled LNA is not able to diffuse into the bleached area in cells with ASOs delivered under gymnotic conditions, and this supports the general interpretation that cellular uptake of LNAs is predominantly by endocytosis. (For color version of this figure, the reader is referred to the online version of this chapter.)

(Flu-LNA) was transfected into cells, the majority of the fluorescence accumulated in the nucleus. When the nucleus-located Flu-LNA was laser photo bleached partial restoration of the fluorescence was noted with time. In contrast to this, when the part of Flu-LNAs concentrated in perinuclear bodies was bleached, no restoration of fluorescence was observed (Figure 2.3). When Flu-LNA was taken up by cells after unassisted delivery, the majority of the fluorescence was found in cytoplasmic-confined bodies. After photo bleaching of these loci the fluorescence was not restored. It is reasonable to assume that reoccurrence of fluorescence in the transfected nucleus reflects that free and intact Flu-LNA is able to diffuse into the bleached area. The static nature of the Flu-LNA obtained during unassisted conditions supports the general interpretation that cellular uptake of LNAs is predominantly by endocytosis. Surprisingly, this uptake pattern was found to be associated with specific and robust antisense activity, measured at both the mRNA and protein level (Stein et al., 2010). The phenomenon was named "Gymnosis" after Greek "gymnos" or "naked," and has now been demonstrated in numerous cellular-based assays. Antisense activity is dose dependent and IC_{50} values are typically obtained at concentrations in the range of 1–25 µM. Gymnotic activity is highly cell line and ON specific. Some cell lines are very sensitive in which a panel of LNAs will exhibit IC_{50} values in the 50–500 nM range. For the most sensitive cell lines, IC_{50}'s may be obtained after 1 day. The vast majority of cell lines tested are gymnotically responsive and only in very few cases have activity not been demonstrated. Not only adherent cell lines are active in gymnosis. Suspension cell lines, such as leukemic cells, are active in much the same way. This is particularly interesting since these cell lines are known to be almost impossible to transfect and before gymnosis was demonstrated not possible to antisense. Furthermore, primary cells harvested directly from live animals are also active under gymnotic conditions.

It is still unknown by which uptake mechanism LNAs are producing antisense activity. Several different theories have been proposed about the ON uptake issue, and they are in the process of being investigated (Alam et al., 2010; Koller et al., 2011). The photo bleaching experiment shown here illustrates that the majority of LNAs are endocytosed and thus internalized by a macropinocytosis (MP)-like process (Reyes-Reyes et al., 2010). It has been shown that MP can be stimulated by PS (Overhoff & Sczakiel, 2005; Rockwell et al., 1997), and this could be one part of the explanation why phosphorothioation is so important for the antisense activity of LNAs. However, this uptake pathway will predominantly internalize LNAs in an "inactive" state/compartment. To "activate" LNAs, they must be released. One possible way could be if the endosomes, during their intracellular trafficking, leak out or by some other mechanism shed the LNAs into the cytoplasm. It is also possible that other uptake mechanisms are involved to achieve the actual antisense effect. It has been shown that nucleic acid channels exist in rat kidney and brain cells (Shi, Gounko, Wang, Ronken, & Hoekstra, 2007), but it has not been demonstrated that this is a general uptake mechanism for other cell types. Instead of "a single ASO channel" mechanism we propose that a complementary uptake mechanism, other than MP, primarily driven by surface/membrane protein binding of LNAs may internalize active antisense compounds. LNAs exhibit a high degree of property heterogeneity also when it comes to protein binding and therefore different LNAs will exhibit a differentiated cell membrane protein binding. When the cell membrane proteins are internalized, the bound LNAs can be released. This proposed mechanism will exhibit cell type and compound differentiated uptake pattern, in line with what is seen in Figure 2.4A and B. Further examination of this is in progress.

3.1.2.2 Comparing cellular uptake and antisense efficiency

A typical gymnotic experiment is illustrated in Figure 2.4A where RNA downregulation is illustrated for two targets. Four LNAs specific for Target 1 and two specific for Target 2 are shown. The LNAs were designed as three sequences, of which each was used unconjugated or conjugated with fluorescein (5920 = Flu-5612; 5921 = Flu-5613; 6919 = Flu-4093). For the six LNAs, the amount in the lysate was calculated and related to the total cell count. In this way, the total number of LNA molecules accumulated in a single cell was calculated. The accumulation was, as expected, related to assay concentration (Figure 2.4B), but the total number of LNA molecules accumulated varied considerably across the compounds. However, the Flu-LNAs were generally taken up better as compared to the unlabeled LNAs and the highest accumulation was observed for 5921 at 25 µM, having 6.5 million LNA molecules per cell. The variance was most pronounced at the lowest concentration (1 µM) exhibiting an uptake ratio of 1:12 between 5613 and 6919. Across the compounds fluorescein-conjugation tended to produce increased uptake, but interestingly that did not correlate with increased activity. Thus, the most potent compound 4093 with 250,000 molecules per cell (1 µM) produced an 80% downregulation of Target 2, but 6919 (Flu-4093) reduced the target by

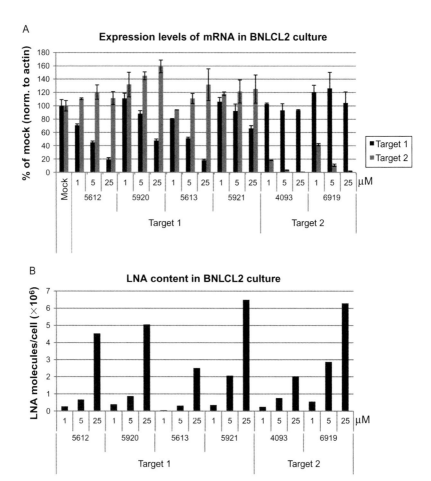

Figure 2.4 RNA downregulation for two targets after gymnotic ASO delivery in cell culture. Expression data for two RNA targets in BNLCL2 cells 3 days after ASO-treatment. Four ASOs (two sequences with and without fluorescence labeling) specific for Target 1 and two ASOs (one sequence with and without Flu-label) specific for Target 2 were added at concentrations of 1, 5, and 25 μM. (A) RNA downregulation against the specific as well as the control target (the other target used as specificity marker) is shown (5920 = Flu-5612; 5921 = Flu-5613; 6919 = Flu-4093). (B) The amount of LNA ONs in the BNLCL2 cell lysates as shown in A was determined by ELISA and related to the total number of cells after incubation to give the total number of LNA molecules accumulated in a single cell.

only 60%, despite the fact that more than the double amount was taken up. These data illustrate that uptake is a compound-specific parameter and although conjugation may favor cellular uptake, it may equally well disfavor the antisense effect.

Gymnosis is now becoming an established technique for *in vitro* antisense drug discovery. It represents an important improvement for early stage screening assays and is believed to get significant impact on the antisense/ RNAi field. Gymnosis avoids introduction of massive amounts of catonic lipids, removing the cellular responses the lipids produce in their own right (Moghimi et al., 2005; Symonds et al., 2005). The viability of cells *in vitro* is much improved and false positive responses imposed by the lipids are eliminated. Most importantly, silencing under gymnotic conditions mimics more closely the *in vivo* setting, and therefore gymnosis creates a higher probability for that data obtained *in vitro* will predict *in vivo* outcome. Thus, *in vitro* systems are becoming better surrogate models for *in vivo* outcome.

From a practical drug development point of view, gymnosis has proven that there is no need to develop complex delivery systems to obtain cellular uptake *in vivo*, a fact that greatly reduces the complexity in developing LNAs as drugs.

3.2. siLNA: Improving siRNA performance by LNA modification

Synthetic siRNAs have truly revolutionized functional genomics in mammalian cell cultures due to their reliability, efficiency and ease of use, but seem ill-suited as a potential RNA therapeutic in a chemically nonmodified form. Consequently, researchers have now for a decade incorporated chemically modified nt analogues into siRNAs with the prospect of enhancing their activity, specificity, nuclease resistance and reducing potential immunogenicity (reviewed in Bramsen & Kjems, 2012). LNA was among the first modification types employed to engineer siRNAs (Braasch et al., 2003; Elmén et al., 2005) and has since proven capable of improving many aspects of siRNA performance (Figure 2.5).

3.2.1 Tolerance for LNA modification

A priori, LNA is expected to be well tolerated in siRNA design as it will conformationally lock siRNA and guide strand–target duplexes in a RNA-like A-type helical configuration (Koshkin et al., 1998) essential to the RNAi process (Chiu & Rana, 2002, 2003). Indeed, single RNA to LNA exchanges are found by systematic empirical testing to be well tolerated at most positions of the siRNA, with notable exceptions. These are the guide strand 5′-terminal base and positions around the mRNA cleavage site at pos. 10, 12, and 14 (Elmén et al., 2005). It was early established that moderate LNA modification of either strand of the siRNA can significantly enhance the T_{m} of the siRNA duplex with little negative impact on its activity

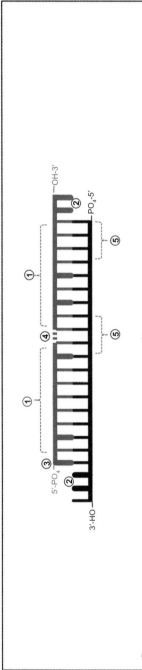

① **Enhancing thermostability by passenger strand mod.**
Moderate LNA-mod. of the passenger strand will enhance siRNA thermostability with little impact on siRNA activity (Braasch et al. 2003), dramatically enhance siRNA nuclease resistance (Elmen et al. 2005; Bramsen et al. 2007, 2009), reduce siRNA immunogenicity (Hornung et al. 2005; Goodchild et al. 2009; Schyth et al. 2012) and silencing specificity by reducing the contribution of the passenger strand to RNAi.

② **Optimizing strand selection by 3′ overhang mod.**
Asymmetrically LNA-mod. 3′ overhangs can favor guide strand selection during RISC loading to enhance siRNA activity and specificity (Bramsen et al. 2009) and simultaneously enhance siRNA nuclease resistance (Elmen et al. 2005; Mook et al. 2007, 2010; Bramsen et al. 2009).

③ **Optimizing siRNA thermo-assymmetry**
LNA-mod. of the 5′-terminal position of the passenger strand will favor guide strand selection and enhance siRNA activity and specificity (Elmen et al. 2005).

④ **Building novel siRNA designs**
The LNA-stabilized sisiRNA design uses two short passenger strands devoid of RNAi activity to enhance silencing specificity, the tolerance for higher (LNA) modification levels and will enhance nuclease stability and reduce immunogenicity as compared to unmod. siRNAs (Bramsen et al. 2007).

⑤ **Preserving guide strand activity**
Single RNA to LNA substitutions are well tolerated in the guide strand except at its 5′ end and at positions opposing the RNA cleavage site (Elmen et al. 2005). Typically, the guide strand is modified only in its 3′ overhang in siLNA designs.

Figure 2.5 A schematic summary of the different aspects where LNA have capacity to improve siRNA performance.

(Braasch et al., 2003; Elmén et al., 2005; also refer to Shen et al., 2011 for a model to calculate the T_m of modified siRNA). However, excessive LNA modification, especially including the seed region (pos. 2–7/8) of the guide strand, impairs siRNA activity (Braasch et al., 2003; Bramsen et al., 2007; Elmén et al., 2005) and seems to prevent efficient siRNA interactions with RNAi proteins (Braasch et al., 2003). Still, the effect of excessive LNA modification can be counteracted by introducing compensating destabilizing modifications or designs; Especially, combining LNA and the highly destabilizing modification unlocked nucleic acid (UNA) or the use of LNA-modified small internally segmented interfering RNAs (sisiRNA), harboring two short passenger strands, have proven highly successful. This maximizes LNA-modification levels, while preserving or even enhancing siRNA functionality and stability (Bramsen et al., 2007; Laursen et al., 2010). It is also notable that the integrity of the sisiRNA design fully relies on LNA modification (Bramsen et al., 2007) and the ongoing efforts to integrate siRNAs in self-assembling DNA/RNA origami structures will expectably benefit from LNA modification in a similar fashion (Afonin et al., 2011).

3.2.2 Enhancing siRNA activity and specificity by LNA modification

A natural priority in siRNA design is to ensure maximal potency for efficient silencing at low concentrations *in vivo*, where siRNA delivery is often limiting. LNA has been successfully used to enhance siRNA activity, primarily by ensuring the preferential loading of the intended guide strand into the RNA-induced silencing complex (RISC). Here, the siRNA strand having its 5′-end embedded at the thermodynamically less stable duplex end will preferentially be incorporated in the active RISC whereas the other strand is cleaved and degraded (Khvorova, Reynolds, & Jayasena, 2003; Matranga, Tomari, Shin, Bartel, & Zamore, 2005; Rand, Petersen, Du, & Wang, 2005; Schwarz et al., 2003). An optimal thermodynamic stability difference between duplex ends is therefore recommended to favor loading of the guide strand (Shabalina, Spiridonov, & Ogurtsov, 2006) and it is important to note that this will not only maximize siRNA potency but also simultaneously reduce the contribution of the passenger strand to unwarranted silencing (i.e., off targeting). One strategy to optimize strand selection is to introduce LNA modifications at the 5′-terminal of the passenger strand, which renders the siRNA duplex thermodynamically highly asymmetric and favors guide strand loading (Elmén et al., 2005). This strategy is particularly attractive if target sequences cannot be freely chosen, for example, when targeting a highly accessible target site of an otherwise highly structured mRNAs. In this regard, the combination of LNA and the destabilizing UNA-modification has proven particularly potent in generating superior thermodynamic asymmetry for efficient and specific silencing (Laursen et al., 2010; Werk et al., 2010). As the strand selection by RISC is also influenced by the structure of the siRNA 3′-overhangs, the use of asymmetric overhangs, that is, removing the 2 nt overhang from the passenger strand, will favor loading of the opposing guide strand and enhance silencing activity and specificity (Sano et al., 2008). Similarly, asymmetric chemical modification of 3′-overhangs has been shown to influence strand selection; In particular, a 3-nt LNA-modified 3′-overhang (5′-$T_{LNA}C_{LNA}$U-3′) was suggested to be favored during RISC loading, whereas a shorter 2-nt LNA-modified 3′-overhang (5′-$T_{LNA}C_{LNA}$) is slightly disfavored. Utilizing these LNA-overhangs in the guide strand and passenger strand of the siRNA, respectively, can hereby enhance silencing activity and specificity regardless of the sequence of the siRNA stem (Bramsen et al., 2009). Finally, thermostabilization of the siRNA duplex through moderate passenger strand LNA modification, a popular approach to enhance nuclease resistance (Braasch et al., 2003; Bramsen et al., 2007, 2009; Dutkiewicz et al., 2008; Elmén et al., 2005; Mook et al., 2010), will also enhance silencing specificity by rendering the passenger strand less functional. In contrast, the strengthening of guide strand–target hybridization through LNA modification of the guide strand does not clearly translate into higher silencing activities (Braasch et al., 2003) as described for ASO–target interactions (Wahlestedt et al., 2000), nor facilitate better silencing of less accessible target sites for typical siRNAs (J.B. Bramsen, unpublished observation). The latter agrees with the observation that miRNA–target hybridization energy is a poor predictor of miRNA silencing activity *per se*, whereas target site accessibility (i.e., the probability of not being intramolecularly base paired) within the structured mRNA seems more important (Marin & Vanicek, 2011). Therefore, guide strands are typically LNA-modified only in their 3′-overhangs to enhance nuclease resistance while preserving silencing activity.

3.2.3 Enhancing siRNA nuclease resistance by LNA modification
Double-stranded siRNAs are much more stable in extracellular environment than their single-stranded counterparts (Bertrand et al., 2002), yet they are still degraded within minutes, for example, in blood serum preparation. This suggests that chemical stabilization is needed during delivery of naked siRNAs *in vivo* (Soutschek et al., 2004) in line with many siRNA drug

candidates being heavily modified (Burnett, Rossi, & Tiemann, 2011). Moreover, chemical stabilization of siRNAs is suggested to enhance the duration of silencing (Chiu & Rana, 2003; Collingwood et al., 2008; Dowler et al., 2006; Takahashi et al., 2012; Volkov et al., 2009), hereby, potentially allowing less frequent siRNA dosing for persistent gene silencing. In many extracellular fluids, siRNA degradation is primarily mediated by endoribonucleases that cleave siRNAs at certain single-stranded dinucleotide motifs presumably exposed by random thermal fluctuations (Sorrentino, 1998). Therefore, chemical modification at these specific positions, typically $2'$-O-Methyl ($2'$Ome) or $2'$-flouro modification of CA and UA dinucleotide motifs, can efficiently enhance siRNA nuclease resistances (Hong et al., 2010; Volkov et al., 2009) and LNA modification will expectably offer similar benefits. More uniquely, the thermostabilizing properties of LNA also allow significantly enhanced nuclease resistance in a more sequence-independent fashion, since moderate enhancement of siRNA thermostability will make the exposure of single-stranded CA and UA motifs to RNase activities in the solvent less likely. Typically, 4–6 modifications are distributed in the passenger strand in order not to reduce the activity of the guide strand and this can lead to a dramatic increase in siRNA nuclease resistance as evaluated in serum upon intravenous injection in mice (Bramsen et al., 2007, 2009; Gao et al., 2009). Also, LNA modification of the $3'$-overhang will not influence silencing activities but significantly enhance siRNA nuclease resistance (Elmén et al., 2005).

3.2.4 Reducing siRNA immunogenicity by LNA modification

It has become clear that unmodified, synthetic siRNAs, despite their structural mimicry of endogenous siRNA/miRNA species, can trigger innate immune responses depending on their structure, sequence, target cell type, entry route, and concentration (Hornung et al., 2005; Judge et al., 2005; Reynolds et al., 2006; Sioud, 2005; Sledz, Holko, de Veer, Silverman, & Williams, 2003). Most prominently, siRNA is detected by members of the transmembrane Toll-like receptor (TLR) family, in particular TLR3 and TLR7/8, which are expressed primarily in immune cells but also in many epi- and endothelial linings of the body (Alexopoulou, Holt, Medzhitov, & Flavell, 2001). Crucial to siRNA recognition are the endosomal TLR7/8, which specifically recognize single-stranded GU or U-rich motifs exposed from the siRNA duplex such as GUCCUUCAA (Judge et al., 2005), UGUGU (Sioud, 2006), UGGC (Fedorov et al., 2006), UCA (Jurk et al., 2011), GU (Diebold, Kaisho, Hemmi, Akira, &

Reis e Sousa, 2004; Heil et al., 2004), and AU (Forsbach et al., 2008). Whereas TLR7/8 responses can be reduced by the modification of the particular immune-stimulatory sequence motif using, for example, $2'$OMe, $2'$-fluoro, and LNA (Diebold et al., 2006; Eberle et al., 2008; Heil et al., 2004; Hornung et al., 2005), the thermostabilizing LNA can reduce the exposure of such single-stranded, immunostimulatory motifs to TLR7/8 in a strategy similar to reducing RNase degradation of exposed siRNA single strands as described above (Hornung et al., 2005). Indeed, Goodchild and colleagues found a significant correlation between the free energy of hybridization of the siRNA duplex and its immunostimulatory activity via TLR7/8, and that introducing the so-called LNA clamps in the non-stimulatory siRNA strand would reduce siRNA immunogenicity upon DOTAP-transfection into human peripheral blood mononuclear cells (Goodchild et al., 2009). This concept of using general thermostabilization of siRNA by LNA modification of only one siRNA strand, typically the passenger strand, to reduce siRNA immunogenicity is confirmed *in vivo* by Schyth and colleagues upon injection of DOTAP-formulated, LNA-modified siRNAs in rainbow trout, where a significant correlation between the siRNA duplex T_m, modulated by varying the levels of LNA modification, and immunogenicity was reported (Schyth et al., 2012).

3.2.5 Applying siLNA in vivo
LNA modification is still not among the most popular modification types for siRNA application *in vivo* (as opposed to the very popular LNA-modified ASO designs), although it offers several unique advantages to siRNAs design in general. This likely reflects the early successes of siRNAs modified by more readily available modification types such as $2'$OMe, $2'$-fluoro and PS (Morrissey et al., 2005; Soutschek et al., 2004) and the heavy focus on formulating (i.e., protecting) siRNA in various carrier systems to facilitate delivery. Still, the significant increase in siRNA serum stability conferred by LNA modification seems immediately beneficial for applications of unformulated siRNA. Even for formulated siRNAs LNA modifications may offer advantages in term of prolonged silencing longevity and reduced immunogenicity. In mouse serum, unmodified siRNA is completely degraded within 5 h but introduction of two LNA nts in each of the $3'$-overhangs of the siRNA duplex increases the serum half-life to over 48 h. Further LNA modification of the sense strand provides siRNAs that show no signs of degradation after 96 h of incubation (Mook, Baas, de Wissel, & Fluiter, 2007). LNA-stabilization of siRNA can indeed enhance

their survival in the bloodstream upon intravenous injection in mice as compared to unmodified siRNA (Gao et al., 2009; Laursen et al., 2010). LNA-stabilized, naked siRNAs have also been successfully used to reduce eGFP expression in the lung of eGFP expressing mice upon intravenous injection; however, nonmodified siRNAs were not included for comparison (Glud et al., 2009). Similarly, Mook and colleagues evaluated the efficacy of unformulated LNA-modified siRNA on silencing of eGFP in a mouse xenograft cancer model upon tail vein injection or using osmotic micropumps; both delivery methods produced a 50% KD of eGFP expression in tumor cells. The slight reduction in siRNA activity observed in cell culture upon 3′-overhang LNA modification seemed to be compensated for by an enhanced stability, hereby showing clear benefit of LNA modification *in vivo*. Indeed, in a subsequent study using a similar setup, Mook and colleagues reported that LNA-modified siRNA and sisiRNA design generated a 50% reduction in eGFP expression whereas an unmodified siRNA was nonfunctional (Mook et al., 2007).

3.3. LNA and splice-switching ONs

3.3.1 Splice switching

Mature mRNA is generated by removal of the noncoding introns in a very precise way. This is directed by a large number of proteins and small nuclear RNAs in the "spliceosome," for a review see Wahl, Will, and Luhrmann (2009). Splicing is controlled by highly conserved sequences around the 5′- and 3′-splice junctions, in combination with silencer and enhancer sequences found both in intron and exon sequences. During the past decade, ASOs, which bind to pre-mRNA and block-specific signal-sequences, have been used to modify the splicing patterns. This can be either by inducing skipping or enhancing inclusion of a specific exon, as illustrated in Figure 2.6, and has become a promising strategy for gene therapy of monogenic disorders, but also in the treatment of inflammatory diseases (Kole, Krainer, & Altman, 2012).

One of the first reports, and proof-of-principle on the usage of splice-switching ONs (SSOs) to restore protein expression, was based on the β-globin gene (Dominski & Kole, 1993). Kole and his group also established a model system, which has been frequently used to compare different chemistries and delivery methods for SSOs (Kang, Cho, & Kole, 1998). The splice-modulating approach, to either induce exon skipping or promote exon inclusion, has since then been investigated for treatment of a number of monogenic diseases, for example, cystic fibrosis (CF), duchenne muscular

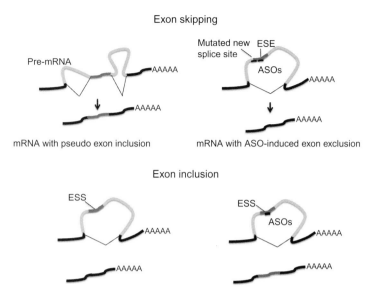

Figure 2.6 A schematic illustration on how ASOs can influence pre-mRNA splicing events. Exon skipping: a mutated pre-mRNA is spliced to include the unwanted pseudoexon. By adding ASOs directed toward the mutated splice site or toward exon splice enhancer (ESE) signals in the pre-mRNA the splicing is influenced to exclude the unwanted exon. Exon inclusion: a pre-mRNA with poor inclusion of a certain exon, due to existing exon splice silencing (ESS) signals in the sequence, is spliced to a shorter mRNA. Blocking the ESS signals with ASOs can promote the inclusion of the exon in the mature mRNA. (For color version of this figure, the reader is referred to the online version of this chapter.)

dystrophy (DMD), spinal muscular atrophy (SMA), and different forms of cancer. For recent reviews on splice-switching therapies and SSO delivery, see (El Andaloussi, Hammond, Mager, & Wood, 2012; Kole et al., 2012; Spitali & Aartsma-Rus, 2012).

SSOs must hybridize efficiently and with high specificity to pre-mRNA in order to block the access of different splicing factors, and this without risking to activate RNase H. Different chemistries have also been used to enhance the binding affinity as well as serum stability, such as phosphorodiamidate morpholino (PMO) (Stirchak, Summerton, & Weller, 1989), 2′OMe PS (Shibahara et al., 1989), tri-cyclo-DNA (Renneberg, Bouliong, Reber, Schumperli, & Leumann, 2002; Renneberg & Leumann, 2002), LNA (Braasch et al., 2002; Kumar et al., 1998; Obika et al., 1998), and peptide nucleic acid (PNA) (Larsen, Bentin, & Nielsen, 1999; Nielsen, Egholm, Berg, & Buchardt, 1991). Among them, both

$2'$Ome and PMO SSOs have already reached clinical trials for treatment of DMD, for recent reports on the clinical usage, see (Cirak et al., 2011; Goemans et al., 2011). For comparisons of SSOs of different chemistries see (Aartsma-Rus et al., 2004; Laufer, Recke, Veldhoen, Trampe, & Restle, 2009), and an extensive overview on how to design SSOs is present in (Aartsma-Rus, 2012). In the next section, we will concentrate on the effects of LNA in SSOs.

3.3.2 LNA in SSOs

Since LNAs do not activate RNase H, they can safely be spiked into SSOs without the risk for pre-mRNA degradation. Moreover, the high affinity for hybridization to complementary RNA allows the use of shortened ONs. On the other hand, while short ONs can be beneficial from the uptake point of view, a very short ON with LNA in essentially all positions, will have a much higher number of complementary off-target sites, with capacity to consume the ON and, thus, reduce or prevent the correction. In the same way, longer ONs with intermediate to high LNA content can bind to off-target sites with partial complementarity. This has also been reported for long $2'$OMe–LNA ONs with relatively high LNA contents (33–50%), showing decreased specificity against mismatched targets (Guterstam et al., 2008). Ittig et al. compared LNA with tri-cyclo-DNA ONs in a cyclophilin A splicing assay (Ittig, Liu, Renneberg, Schumperli, & Leumann, 2004). They found that despite a higher T_m against the target RNA, when transfected with lipofectamine into HeLa cells, the LNA ONs were less efficient than the corresponding tri-cyclo-DNAs. The difference in backbone influenced the nuclear distribution, with a higher concentration of LNAs in what was assumed to be the nucleoli. In this study, the 9-mer LNA did not give any splicing effects, while at 0.2 µM concentration, both the 13- and the 15-mer LNAs reduced the full-length mRNA by 60%.

In 2006, the use of evenly distributed LNA monomers in a 16-mer DNA ON with PS backbone was shown to efficiently correct the splicing of the EGFP reporter-gene *in vivo* in a transgenic mouse model (Roberts et al., 2006). After four daily intraperitoneal (i.p.) injections, the strongest splice-correction was seen in the liver, small intestine, and colon, while hardly any effect was found in the heart. Even after high-dose oral delivery at 50 mg SSO/kg, a very weak but clear effect could be detected in the intestine and the liver.

Tumor necrosis factor-α (TNF-α) is a key cytokine in many inflammatory diseases. Graziewicz et al. have used LNA-containing SSOs to redirect the splicing of the TNF-α receptor 2 (TNFR2) toward the alternatively spliced secreted form, missing the transmembrane part (Graziewicz et al., 2008). This naturally occurring version of TNFR2 acts as an efficient antagonist for TNF-α. By increasing the proportion of mRNAs spliced to this isoform, an antiinflammatory, nonimmunogenic molecule will be constantly and systemically provided. Daily i.p. injections for 5–10 days with SSOs reduced the severity of the inflammation in both a hepatitis and an arthritis mouse model.

Since LNA-containing ASOs can also be taken up in lymphocytes, using gymnosis (see above), this modification has also been used in SSOs for treatment of pseudoexon mutations in the B-cell linage (Bestas et al., 2012). A battery of LNA-2'OMe-mixed SSOs was tested and two of the most promising 15-mer ONs have been used to correct expression from the gene encoding Brutons tyrosine kinase (BTK). Expression of BTK was shown both in B-cells from a humanized *BTK*-transgenic mouse and in monocytes from a patient suffering of the disease X-linked agammaglobulinemia after both electroporation and gymnotic transfections.

3.4. LNA in antigene reagents

Despite the large number of chemically modified nucleic acid analogues that have been developed during the past two decades, few of the new chemistries have become characterized when it comes to their performance as antigene reagents. Antigene ONs can act via blocking the binding of transcription factors or by stalling the RNA polymerases. These ONs can be divided into different categories due to the mode of binding to the target DNA duplex, that is, (i) as triplex-forming ONs (TFOs) inserted in the major grove, (ii) by strand invasion and WC hybridization, and (iii) as a clamp-type ON combining these two modes of interactions. The requirements of high affinity hybridization in combination with strong sequence specificity under physiological salt and pH conditions have reduced the number of modifications that have been found interesting as antigene ONs in living cells.

3.4.1 LNA in triplex-forming ONs

Most TFOs are either pyrimidine- or purine-ONs. They target stretches with homopurine–homopyrimidine DNA duplexes where they bind in the major grove with high sequence specificity via Hoogsteen and reverse Hoogsteen interactions, respectively. Purine TFOs thus bind antiparallel to the purine-strand of the target-DNA duplex while pyrimidine TFOs bind in a parallel setting. In addition, mixed GT-containing TFOs exist, which can hybridize in both orientations. For an early review on triplex formations, see Frank-Kamenetskii and Mirkin (1995). Since G-rich purine-TFOs frequently aggregate under physiological ion conditions, most studies so far have been dealing with pyrimidine TFOs. Still, also the pyrimidine TFOs have limitations, since the cytosine monomers need to be protonated in the N3 position in order to give a stable contribution to the triplex. Thus, when unmodified cytosines are incorporated, the pH must be below 6, that is, well below what is found in the cell nucleus. Thus, many efforts have been made to find DNA and RNA analogues, which can form stable triplexes under intranuclear conditions.

3.4.1.1 LNA-modified pyrimidine TFOs

LNA is a very interesting modification in this regard. The conformationally restricted backbone of LNA-containing ONs preorganizes the TFOs in a favorable conformation for hybridization toward a dsDNA target. Already in 2001, Imanishi and coworkers demonstrated the capacity of BNA (LNA)-containing pyrimidine-ONs to form stable triplexes with a dsDNA target at the physiological pH 7.2 (Obika et al., 2001; Torigoe, Hari, Sekiguchi, Obika, & Imanishi, 2001). Short, 14-mer TFOs with every second or every third base as LNA were shown to form stable triplexes, while fully LNA-substituted ONs did not (Obika et al., 2001). Notably, only TFOs containing methylated cytosines were stable in neutral pH at temperatures of 36–37 °C, as demonstrated both by T_m measurements and DNase protection. Commercially available LNA-C amidites are normally methylated, which then, although to a lower degree, compensates for the loss of protonation under physiological pH. Despite this, it is important to also include a number of LNA-thymidines in the TFO to get stable triplex formation under these conditions (Sun et al., 2004).

When it comes to LNA-modified TFOs, it was early found that the increased binding constants at neutral pH was mainly due to the considerable decrease in k_{off} (Torigoe et al., 2001). During the years, a number of LNA modifications have been analyzed for their capacity to enhance or stabilize triplex formation at neutral pH. The 2′-amino-LNA (Singh, Kumar, & Wengel, 1998b) provides a "handle" through which it is possible to functionalize the LNA by adding different modifications. Although the 2′-amino-LNA by itself destabilizes the triplex as compared to LNA-substituted TFOs, adding a glycyl-group to the amino-LNA stabilized the triplex with an increase in T_m of 1.7–3.5 °C per modification as compared to the original LNA–TFO (Højland, Babu, Bryld, & Wengel, 2007). A more extensive comparison of different $2'$-amino-LNAs and other LNA modifications can also be found in (Højland, Kumar, et al., 2007). Some other modifications placed within the bridge of the LNA sugar, like $2'$-$O,4'$-C-aminomethylene ($2',4'$-BNA^{NC}), were also shown to increase the stability of the formed triplex under physiological pH as compared to the original LNA-modified TFO (Rahman et al., 2008). Again the increase in binding constant was almost exclusively related to a decrease in the off-rate (Torigoe, Sato, & Nagasawa, 2012). Additional modifications of LNA, which stabilize TFO formation as compared to ordinary DNA–TFOs, have been reported, but unfortunately without comparisons with TFOs containing the corresponding unmodified LNAs (Sasaki et al., 2009). Brunet et al. addressed the protonation problem under physiological pH conditions by exchanging a number of consecutive cytosines in the end of an LNA-containing pyrimidine TFO with guanines, forming a TCG-parallel LNA-containing TFO. This reduced the k_{on} to some degree but again decreased the k_{off} considerably as compared to the original pyrimidine-LNA–TFO (Brunet, Alberti, et al., 2005).

Beside the conformational preorganization and the improved hydrogen bonding provided by the cytosine methylation, further stabilization of the triplex can be achieved by improving base stacking. C5-alkynyl-modifications of LNA with C5-ethynyl or C5-propargylamine monomers were reported to increase the T_m of the triplex with between 1.1 and 2.8 °C per modification as compared to the unmodified LNA (Sau et al., 2009). Importantly, the C5-alkynyl-modifications seem to keep, and even improve, the mismatch discrimination against the dsDNA target. Interestingly, alkynyl-modifications, and then especially the C5-propargyl-modification, were also shown to provide increased protection toward the snake venom $3'$-exonuclease.

3.4.1.2 Intracellular effects of LNA-modified pyrimidine TFOs

Both purine and pyrimidine TFOs have been tagged with psoralene (Pso) and used to induce photo-activated DNA repair in living cells. Importantly, in a comparison between a number of TFOs without capacity to form such targeted DNA adducts, a N3$'$P5$'$ phosphoramidate-TFO induced DNA repair in cell extracts, while the corresponding LNA-containing pyrimidine TFO did not. Neither did the LNA-containing TFO increase the recombination rates after transfection into CHO cells (Kalish, Seidman, Weeks, & Glazer, 2005). Still, by pre-hybridizing a reporter plasmid with an

LNA–TFO targeting, a sequence just upstream of a reporter gene, Brunet et al. could show a 50% reduction in the expression of the targeted luciferase-reporter at 2 μM concentration. The corresponding Pso-conjugated TFO displayed almost the same reduction at 0.5 μM (Brunet, Alberti, et al., 2005), while an acridine-conjugated LNA–TFO was as efficient already at 100 nM (Brunet, Corgnali, et al., 2005).

Targeting genomic sequences inside the cell will, however, provide additional constraints on the TFO formation, the chromatin state being one. Using Pso-conjugated pyrimidine-LNA–TFOs directed toward two different genomic sites, triplex formation in the intact chromatin setting was demonstrated (Brunet, Corgnali, Cannata, Perrouault, & Giovannangeli, 2006). These experiments were performed in digitonin-permeabilized mammalian cells. After Pso-cross-linking the TFO formation in cells with different expression profiles was investigated by PCR. After Trichostatin A treatment to induce enhanced transcription, a 2- to 10-fold increase in TFO-mediated modification of targeted genes was noticed. Still, there seems not to be any requirement for gene expression to allow TFO binding, since also silenced reporter genes could be targeted in stably transfected cells (Brunet et al., 2006).

Interestingly, the TFOs with the highest thermal stability are not necessarily the best when it comes to targeting specific sites in the chromatin. Alam et al. used a gene mutation assay to compare the specific *in vivo* targeting with the *in vitro* TFO-hybridization efficiency (Alam et al., 2007). All their TFOs contained a patch of four ENA nts in the 3′-end as well as a 5′ Pso-moiety. With increasing number of LNA modifications in the remaining TFO-sequence they concluded that the TFOs with highest thermal stability lost the capacity to bind to the target gene in the cell context. A possible explanation is that, the higher the LNA content the more "off-target" sites can bind the ON and reduce the concentration at the specific target site. A way to investigate this theory would be to use biotinylated Pso-TFOs, hybridize, crosslink, pull down and sequence the DNA-fragments bound to the TFO.

3.4.2 LNA in double-strand-invading ONs

While TFOs bind to the DNA duplex via hydrogen bonding in the major grove, double-strand invasion (DSI) can occur when an ON displaces one of the strands in the DNA duplex and forms WC based hydrogen bonding with the second strand. Due to the strong stacking interactions the displacement of existing hydrogen bonding in long DNA duplexes, is rather infrequent (Yakovchuk, Protozanova, & Frank-Kamenetskii, 2006). In-crease in temperature or negative super-coiling of the DNA increases the "breathing," the transient base-pair openings that occur in a DNA duplex (Frank-Kamenetskii, 1987; Gueron, Kochoyan, & Leroy, 1987), which will facilitate DSI. This general concept was clearly verified by studies on the effects of temperature and DNA super-coiling for PNA-mediated DSI (Bentin & Nielsen, 1996).

3.4.2.1 LNA-containing clamp ONs

An interesting approach to gene targeting is the formation of clamp-like ONs, which contain a TFO-strand connected to a WC base-pairing strand via a flexible linker. The two strands are targeting the same polypurin stretch. This type of ON was first reported for the nucleic acid analogue PNA and was named bispNA (Egholm et al., 1995), for reviews on PNA see (Lundin, Good, Stromberg, Graslund, & Smith, 2006; Nielsen, 2010).

The first report on an LNA-containing bis-clamp, was made by Hertoghs, Ellis, and Catchpole (2003). They used a 9-mer bisLNA as an anchor to functionalize a plasmid for vaccination. They found the tested bisLNA less efficient than similar bispNA-clamps, although the hybridiza-tion was performed at low salt and low pH, and the anchor was targeting to multiple sites in the plasmid. It is well known that multiple target sites within a few base-pairs distance increase the DSI by cooperative bind-ing (Lundin et al., 2004). Recently, Moreno et al. reported improved bisLNA-constructs consisting of triplex-forming antiparallel pyrimidine-clamp DNA–LNA mixmer ONs. These optimized bisLNAs proved to have DSI capacity under physiological conditions also when targeting a single site (Moreno et al., 2013). Schematic illustrations of different DSI ONs can be found in Figure 2.7. In these bisLNAs, the WC base-pairing part was con-tinuously synthesized with the TFO via a linker consisting of five pyrimi-dines. It is well established that the presence of LNA residues stabilizes duplex as well as triplex formation. Here, it was investigated how different positioning of LNA nts influenced the DSI when LNA residues were inserted in both the WC and the Hoogsteen strand. The N-type LNA sugar pucker adapts the ON to a favorable conformation for pyrimidine triple-helix formation. At the same time, an LNA-containing WC duplex adapts to an A-type, RNA-like, structure (Bruylants, Boccongelli, Snoussi, & Bartik, 2009; Petersen et al., 2002). Thus, as could be expected, the clamp with LNA in both arms displayed the highest T_m. This is in line with the results from a study where a parallel pyrimidine–purine clamp with an

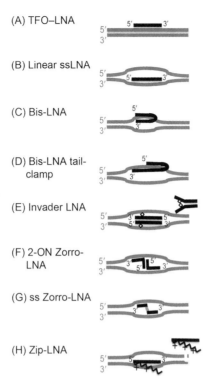

(A) TFO–LNA

(B) Linear ssLNA

(C) Bis-LNA

(D) Bis-LNA tail-clamp

(E) Invader LNA

(F) 2-ON Zorro-LNA

(G) ss Zorro-LNA

(H) Zip-LNA

Figure 2.7 A schematic illustrations on the binding mode for different LNA ONs with assumed antigene capacity. *TFO–LNAs* bind to duplex DNA in the major grove, via Hoogsteen or reverse Hoogsteen interactions. A TFO is restricted to bind to a polypurine stretches in the DNA duplex. *BisLNA* and *bisLNA tail-clamps* bind via a combination of Watson–Crick (WC) base pairing and TFO interactions while the rest of the ONs bind by normal WC interaction after strand invasion of the DNA duplex. *Invader-LNAs* are modified with two 2'-*N*-(pyren-1-yl)methyl-2'-amino-α-L-LNA monomers, one on each strand in adjacent base pairs, to prevent ON–ON hybridization. Thus, they can hybridize to overlapping sites on the two opposite strands of the DNA duplex, while *Zorro-LNA* and single-stranded (*ss*)*Zorro-LNAs* avoid ON–ON interactions by targeting adjacent sites on the two DNA strands. Zip-nucleic acids are conjugated with oligospermine tails in order reduce the charge and to facilitate cellular uptake of the ON. (For color version of this figure, the reader is referred to the online version of this chapter.)

LNA-containing pyrimidine-TFO-arm formed more stable triplexes with RNA in the pyrimidine WC strand than when the duplex was all DNA (Alvira & Eritja, 2010). If the LNA nts were placed in the same base triade, or if they were placed alternating in the TFO and WC strand, did not influence the DSI (Moreno et al., 2013). The most efficient strand invasion was found when using a "tail-clamp," (Figure 2.7D). The tail of the tail-clamp was shown to be important for efficient DSI under physiological conditions, while the TFO-part was important for stabilization of the complex. Without the TFO-part, the construct was "kicked out" when the target-DNA duplex was relaxing.

3.4.2.2 Invader-LNA

DNA duplexes modified with two $2'$-N-(pyren-1-yl)methyl-$2'$-amino-α-L-LNA monomers, one on each strand in adjacent base pairs (Figure 2.7E) are relatively unstable, as compared to when the duplex only contains the modification on one strand. DNA duplex ONs with this "+1 interstrand zipper arrangement" were called "Invader-LNA" (Sau, Kumar, & Hrdlicka, 2010). Like pseudocomplementary PNA (Demidov et al., 2002), the modified LNA monomers are poorly accommodated in the probe duplexes. A single modification increased the thermal affinity toward unmodified complementary DNA with up to $+19.5\,^{\circ}\mathrm{C}$ (Sau et al., 2010). Double-stranded probes with two sequential "hotspots" were efficiently displacing the strands in an isosequential DNA duplex under physiological ion conditions. However, the DSI-efficiency in longer DNA duplexes containing a short single target site remains to be investigated.

3.4.2.3 Zorro-LNA

In 2007, Ge et al. presented a construct, which they named Zorro-LNA. This molecule was created by annealing two short ONs via a complementary linker region to form a reagent that could simultaneously hybridize to adjacent sites on both strands in the DNA duplex, thus forming a Z-shaped DSI molecule (Ge et al., 2007) (Figure 2.7F). By hybridizing Zorro-LNA to multiple sites cloned in the $5'$-untranslated region in a reporter plasmid they showed that a minimum of two such molecules was required in order to efficiently block reporter gene expression. In addition, microinjection of Zorro-LNA could induce gene silencing of the reporter gene in stably transfected NIH3T3 cells. Later, it was also reported that Zorro-LNA has capacity to block RNA-polymerase III as well (Ge et al., 2008). An improvement in the Zorro-LNA design was presented by Zaghloul et al. (2011). New reverse amidites for LNA synthesis made it possible to synthesize Zorro-LNAs as a single ON with the two arms in opposite directions, creating a $3'$-$5'$-$5'$-$3'$ ON, designated ssZorro (Figure 2.7G). In addition, in order to reduce the size as well as to facilitate the design of different ssZorro-LNAs, the LNA-based linker was exchanged for other linkers. A number of different chemical linkers were used and the different ssZorro-LNAs tested for DSI in plasmids. Short hydrophobic linkers of 3 or12 carbon atoms in length were performing better than long flexible or hydrophobic linkers like a $2 \times$ hexaethylene glycol or $3 \times C12$ linker respectively. The shortest linker, consisting of a single phosphate group, was almost as good in DSI as the original double-stranded base-paired LNA linker (Zaghloul et al., 2011).

The use of Zorro-LNA as an efficient antigene reagent was also reported by Ling, Hou, and Hoffman (2011). They used Zorro-LNA to block transcription of the neurofibromatosis gene 1 (*NF1*) in fibroblast cell lines. They targeted a sequence close to a site for the CCCTC-binding factor (CTCF), and demonstrated gene-specific alterations in the long range DNA interaction, with gene-specific alterations both in CTCF and RNA-pol II binding. Recently, Smith and coworkers have also presented a set of Zorro-LNAs with capacity to block *Huntingtin* mRNA expression (Zaghloul et al., 2012).

3.4.2.4 "Linear" antigene LNA ONs

Although "linear" DSI-LNAs are easily "kicked out" by the competing DNA strand in relaxing duplex DNA, the capacity of an LNA-modified ON to break up a short DNA-hairpin loop (Kaur, Wengel, & Maiti, 2007) indicated possibilities also for such molecules to be active as antigene compounds. Corey and coworkers reported on the capacity of "linear" antigene LNA–DNA mixmer ONs to downregulate the expression of the progesterone receptor (PR) and the androgen receptor (AR) in MCF-7 cells (Beane, Gabillet, Montaillier, Arar, & Corey, 2008; Beane et al., 2007). In the first report, they found the strongest reduction in protein expression (50%), when targeting sites close to the transcription start site. The length of the LNA ONs was important, and only the longer ONs of at least 19 bases were able to induce substantial gene silencing (Beane et al., 2007). In the follow-up study, they used LNA- or ENA-modified antigene ONs targeting the *AR* gene or the two different promoters in the PR gene. The ON targeting the *AR* and the *PR-B* transcription start sites was the most efficient and this could be correlated to a reduced co-precipitation of RNA polymerase II. For the *PR-B*, there was also a reduced occupancy of the SP1 transcription factor at the target site present close to the ON recognition site (Beane et al., 2008). This further supports an antigene mode of action.

A further example is the LNA-containing ONs directed toward the hepatitis B virus (HBV) pre-genomic DNA which were transfected into HBV-producing HepG2 human liver cells (Sun et al., 2011). In this study, they found significantly reduced production of HBV antigens after transfection of the sense-ON but only after complexation with lipofectamin. One of the main hurdles for ON-based therapies is the poor cellular uptake. At least in cell cultures, longer negatively charged ONs have considerable difficulties to translocate over the cell membrane, unless disguised in positively charged complexes. The zip-nucleic acid (ZNA) technology in which cationic oligospermine tails are conjugated to the ON (Noir, Kotera, Pons, Remy, & Behr, 2008) provides an alternative to such formulations. In a comparison between ZNA-modified DNA and LNA ONs, Gagnon et al. showed that oligospermine-conjugated antigene LNA against the PR-gene retained capacity to block the expression. Still, the inhibition decreased with increasing oligospermine-tail lengths. Despite this, LNA-modified ZNA-ONs at 500 nM concentration possessed capacity to reduce gene expression (around 40%) without lipid complexation, while the same, but unconjugated, LNA as well as the conjugated DNA ONs did not (Gagnon et al., 2011). Another method to enhance uptake is by using address labels, such as the tri-methylated, m3G-CAP, which selectively enhances uptake into the nucleus (Moreno et al., 2009).

4. LNA PHARMACOLOGY

It is of critical importance in drug development to study the fate of candidate compounds in an organism. Factors such as absorption, disposition, target organ and cell, metabolism, residence time, and excretion have a deep impact on the pharmacological and toxicological effect(s) of the product. During the past years, a number of such studies have been performed for different LNA compounds, and extensive knowledge has been obtained which will be summarized here.

4.1. Pharmacokinetics
4.1.1 Delivery and bioavailability
In most studies, LNA compounds have been used as "naked" PS-modified ONs, since the native PO counterparts are less stable in vivo and PS improve PK properties. LNAs have been administered into many different animal models, such as mice, rats, pigs, sheep, monkeys, chimpanzees as well as into humans, and have been dosed intravenously (i.v.), i.p., subcutaneously (s.c.), orally (p.o.), intrathechally (i.t.), and locally to the brain (Elmen, Lindow, Schutz, et al., 2008; Elmen, Lindow, Silahtaroglu, et al., 2008; Fluiter, Housman, Ten Asbroek, & Baas, 2003; Fluiter, ten Asbroek, et al., 2003;

Lanford et al., 2010; Moschos et al., 2011; Park et al., 2011; Straarup et al., 2010; Wahlestedt et al., 2000). The absorption of tested LNAs after s.c. administration is usually complete with 100% bioavailability (BAV), whereas BAV upon oral delivery is substantially lower (10–15%) (data not shown).

The PK profile of PS-modified LNAs *in vivo* is characterized by rapid initial clearance from plasma that results from the redistribution and uptake by peripheral tissues. Within 24 h after i.v. administration, 95–99% is cleared from the blood. The maximum concentrations in plasma (C_{max}) and area-under-the-plasma-concentration-versus-time-curve (AUCs) increase dose proportionally (Lanford et al., 2010). The volume of distribution of these compounds is more than two orders of magnitude larger than the blood volume, indicating a localization to deeper compartments (peripheral tissues), whereas the total body clearance is low (Hildebrandt-Eriksen et al., 2012). The initial plasma clearance has a half-life of less than 2 h, while the clearance from tissues has a half-life in the order of weeks. This characteristic of PS-modified LNAs has been exploited in nonclinical toxicology studies where doses were administered as infrequently as weekly (Hildebrandt-Eriksen et al., 2012; Lanford et al., 2010). The plasma kinetics of LNAs in monkeys is quite similar to that in man and therefore dose levels scale between these species directly on the basis of body weight and not on the basis of surface area, as exemplified in Figure 2.8.

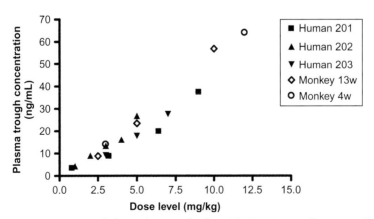

Figure 2.8 Comparison of plasma kinetics for the ASO Miravirsen in human and monkeys. Linearity of exposure (as indicated by plasma area-under-the curve, AUC) versus dose level in mg/kg for humans and monkeys treated with Miravirsen (SPC3649). Human data are from three clinical studies and the monkey data are from two toxicity studies (4-week and 13-week studies).

During their transient time in circulation the LNAs are bound to plasma proteins. The binding capacity varies between compounds, and for any compound also between different animal models. Normally, the protein binding is between 85% and 95% (data not shown). The LNAs bind to a higher degree to rat and monkey plasma proteins (and to human serum albumin) than to mouse plasma proteins. A relatively high plasma protein binding is probably important for a higher uptake in various tissues, since uncharged ONs and the ones containing PO linkages, which bind to plasma protein to a lower degree, are rapidly excreted in the urine (data not shown).

4.1.2 Tissue distribution

Many tissues take up LNAs, although liver and kidney are normally the organs with the highest uptake, as shown by whole body autoradiography of mice administered with tritiated-SPC2996 (16-mer LNA gapmer) (Straarup et al., 2010). The distribution pattern is probably a result of different blood flow in the various organs/tissues. The LNA ONs are uniformly distributed in the liver, although some studies have shown a preference for Kupffer cells (Elmen, Lindow, Schutz, et al., 2008; Obad et al., 2011; Straarup et al., 2010). It was found that 24 h post-dose the accumulation of compound in skin, bone marrow, spleen, uterus, lymph node, and lung was 120%, 80%, 70%, 70%, 50%, and 40% of the accumulation in the liver (Koch & Ørum, 2007; Obad et al., 2011). In the kidney, the compounds are normally concentrated to the cortical region because the LNAs are reabsorbed in proximal tubili cells.

Earlier studies appeared to indicate similar accumulation in the liver of a few but different LNAs. Thus, after a single i.v. dose of 5 mg/kg of a 13-mer LNA compound to mice, about 5 µg/g tissue was detected in the liver (Straarup et al., 2010). Today, we know that the uptake and accumulation in tissues are highly compound/sequence specific. The concentration of different LNAs in, for example, liver could be totally different and ranging from 2 to as much as 70 µg/g liver tissue after a 5 mg/kg dose (Laxton et al., 2011). In general, the half-life of LNA compounds in tissues is long and similar between different tissues, except for the kidney. The kidney half-life may be both shorter and longer than for the other tissues. However, the half-life is specific for a certain compound, and interestingly, the tissue half-life may vary among different species. In mice, the half-life in the liver usually ranges between 4 and 10 days, whereas it is considerably longer in monkeys (2–4 weeks), and even longer in humans with up to 5 weeks. The half-life of the terminal phase in plasma normally equals the tissue half-life. The length of these half-lives indicates extreme stability of LNA compounds *in vivo*. Preliminary studies show degradation only to a minor degree, less than 4%, in plasma samples collected 72 h after dosing (Santaris Pharma).

The activity of a given LNA compound in tissues is directly related to the uptake. However, the specific activity of a compound does not necessarily correlate with its accumulation, that is, the LNA compound with the highest activity may not be the one that accumulates the most (Geary et al., 2009; Moschos et al., 2011; Obad et al., 2011). Moschos et al. showed that inhalation of an LNA compound led to sufficient uptake of the compound to enter the circulation and become redirected to the liver to exert its targeted gene knockdown (Moschos et al., 2011). Since the uptake was lower after inhalation, but the activity was similar, these results could be interpreted as if uptake of the compound in liver after inhalation gives a higher specific activity of the compound. In line with these observations are studies suggesting that the cellular uptake of ASOs is very complex and may occur by several pathways. Some, like MP (*vide supra*) predominantly leaves the LNAs "trapped" in endosomes. Unspecific binding of LNAs to membrane proteins followed by internalization may instead cause intracellular release of LNA as "free" and active ONs. In order to be able to affect the uptake of the LNA compounds in a direction toward higher specific activity, additional work is required to characterize these pathways.

4.2. Pharmacodynamics

Pharmacology of LNA has over the past 12 years been explored in mice, rats, nonhuman primates (NHPs), chimpanzees, and in man. Among these studies, there are numerous examples of pharmacology demonstrated in models of human diseases like different cancers (Corsten et al., 2007; Emmrich, Wang, John, Li, & Putzer, 2009; Fluiter, Frieden, Vreijling, Koch, & Baas, 2005; Fluiter, Frieden, Vreijling, et al., 2005; Fluiter, Housman, et al., 2003; Fluiter, ten Asbroek, et al., 2003; Greenberger et al., 2008; Hansen et al., 2008; Sapra et al., 2010; Zhang et al., 2011), inflammatory disorders (Graziewicz et al., 2008; Worm et al., 2009), dyslipidemia (Lindholm et al., 2012; Moschos et al., 2011; Straarup et al., 2010), disorders of the central nervous system (Wahlestedt et al., 2000), and infectious diseases (Lanford et al., 2010; Laxton et al., 2011).

The first example of a pharmacological LNA *in vivo* was obtained in rats. LNAs against δ-opioid receptor (Wahlestedt et al., 2000) were administered locally to the brain and targeted to interfere with pain regulation. This pioneering work led to a multitude of studies where targets have been reported to be antagonized in more than a dozen different tissues (Elmen, Lindow, Schutz, et al., 2008; Elmen, Lindow, Silahtaroglu, et al., 2008; Koch & Ørum, 2007; Koch et al., 2008; Lanford et al., 2010; Moschos et al., 2011; Obad et al., 2011; Park et al., 2011; Simoes-Wust et al., 2004; Straarup et al., 2010). Most reports have described activity in liver, kidney, adipose tissue, cells of the blood, heart, small intestine, and bone marrow but also a long range of malignant tissues has been examined. A comprehensive review of the past decade's pharmacology work is beyond the scope of this review. In this context, only a few recent advances in pharmacology of LNA antisense activity in diseases related to targets in the liver and in malignant tissue will be reviewed.

4.2.1 Targeting Apob-100

The liver-expressed ApoB-100 was the first target to be silenced with a cholesterol-conjugated siRNA in experimental animals (Soutschek et al., 2004). Since then ApoB has served as "a golden standard" for bench marking activity of ON-mediated silencing *in vivo*. ApoB is also a promising target for treatment of hypercholesterolemia (HoFH) and a new drug application for the ApoB targeting antisense drug Mipomersen has recently been approved by the FDA for the treatment of patients with homozygous familial HoFH. In a recent pre-clinical study, Straarup et al. showed in mice that the potency of ApoB-100 downregulation increased significantly when shortening LNA ON length from a "parent" 16-mer (Straarup et al., 2010). The shorter LNAs (12- and 13-mers, "shortmers") were designed by truncating the 16-mer sequence from either end. The potency increase by the shortmers was more than 10 times compared to the potency of the 16-mer, and in that study a single dose of 0.5 mg/kg was enough to reduce the expression of apoB mRNA about 50% one day after administration. This dose led to a LDL-cholesterol (LDL-C) reduction of 25% from baseline. Biodistribution with ^{35}S-radiolabeling of the 16-mer and the 12-mer showed that both LNA ONs were taken up in a range of tissues at approx. same concentrations, including accumulation in the liver. Thus, the increased potency of the shortmers could not be explained by differences in tissue uptake. Elimination rate or differentiated tissue half-lives were neither the explanation. The authors speculated that intracellular factors may favor the shorter molecules. Another interesting finding was that shortmers were observed to exhibit improved target specificity explained by increased mismatch sensitivity of the shorter sequences. A 13-mer LNA from the mouse study was also tested in NHPs. Following five doses over 4 weeks of 2 mg/kg, the expression of ApoB mRNA was reduced to 50%, corresponding with a similar and specific reduction in LDL-C. The 13-mer was well tolerated in NHPs even after five doses of 32 mg/kg over 4 weeks. The authors introduced the concept of "threshold affinity" where an affinity limit obtained by lengthening the ON dictates how much this key component can contribute in terms of potency and concluded that it is important to balance target affinity against ON length, and that high-potency LNAs are sequence- and design-dependent and will typically be in the range of 12- to 16-mers (Straarup et al., 2010).

4.2.2 Targeting PCSK9

Another liver-expressed target investigated for antisense is proprotein convertase subtilisin/kexin type 9 (PCSK9) (Graham et al., 2007). This target has over the past number of years attracted great interest in hyper-cholesterolemia treatment. Lindholm et al. tested two LNA gapmers (13- and 14-mer) in NHPs against *PCSK9* (Lindholm et al., 2012). The LNA 13-mer reduced LDL-C by up to 50% at a single dose of 10 mg/kg, an effect that lasted for 3 weeks. After multiple doses during 4 weeks the most potent compound produced an 85% reduction in both liver *PCSK9* mRNA and serum PCSK9. The specific reduction in LDL-C of 50% was not associated with a change in HDL-C or glucose levels. A 50% reduction was obtained after the first maintenance dose and this level remained stable throughout the 4-week study. After a total dose of 40 mg/kg (day 30), the liver had accumulated approx. 50 μg/gram liver tissue corresponding with the tissue content found at similar conditions in the ApoB study (*vide supra*). The tissue contents of the *ApoB*- and *PCSK9*-targeting LNAs in both cases exhibited an 85% reduction in the target mRNA. The fact that the target downregulation in these two cases seems to correlate with comparable ON amounts accumulated is likely coincidental, since other studies have shown differentiated PK/Pharmaco-dynamics relationship.

4.2.3 Targeting miR-122

Inhibition of noncoding RNAs, for example, microRNAs, is a target class for which antisense LNA has proven to be highly productive. Elmen et al. reported that SPC3649, a 15-mer mixmer, in NHPs potently and dose dependently inhibited the function of microRNA-122 (miR-122) that is highly expressed in the liver (Elmen, Lindow, Schutz, et al., 2008; Krutzfeldt et al., 2005). MiR-122 is important during embryogenesis and regulates several central pathways such as cholesterol/fatty acid synthesis and iron homeostasis. In addition, miR-122 serves as a crucial host factor and protector for hepatitis C virus (HCV) (Jopling, Yi, Lancaster, Lemon, & Sarnow, 2005; Lanford et al., 2010; Machlin, Sarnow, & Sagan, 2011; Shimakami et al., 2012). Accordingly, antagonizing miR-122 has two read-outs: (1) to lower circulating plasma cholesterol levels (Esau et al., 2006; Krutzfeldt et al., 2005) and (2) to reduce liver and plasma viral titers in HCV-infected individuals (Lanford et al., 2010).

In the work reported by Elmen et al. NHPs were administered in three doses of 1, 3, and 10 mg/kg of SPC3649, respectively, on day 1, 3, and 5. Liver biopsies were examined for the presence of drug and LNA/miR-122 duplex formation (Elmen, Lindow, Schutz, et al., 2008). Following three doses of 3 mg/kg, free miR-122 was eliminated and complete duplex formation was detected. It was also demonstrated by *in situ* hybridization that SPC3649 was distributed throughout the liver tissue. In the high-dose group, the maximum reduction in plasma cholesterol was reached after 3 weeks and was approximately 40%.

Lanford et al. tested SPC3649 against the most prevalent HCV genotype in Western societies (genotype 1) in infected chimpanzees (Lanford et al., 2010). Two groups of two individuals were dosed over 12 weeks, with respectively 1 and 5 mg/kg. Liver biopsies were harvested before and after the dosing period and multiple tissue and plasma-specific parameters were measured. Full LNA/miR-122 duplex formation was observed even in the low dose group at the end of the 12 weeks of treatment. In the high-dose group, the LNA/miR-122 duplex was stably present until 14 weeks after the last dose. In this group, viral titers were reduced by approximately 2.6 logs in plasma, and a bit less in the liver, and cholesterol levels were reduced by 30–44%. The kinetics of cholesterol and virus titer reduction followed each other closely, reaching nadirs at the end of the dosing period. The viral titers remained at the "nadir" level for about 9 weeks and then rebounded slowly over 8 weeks toward baseline levels, at which point free miR-122 was also detected. The study was not designed to follow the LNA/miR-122 duplex formation rate, but according to Elmen, Lindow, Schutz, et al. (2008), it is reasonable to assume that free miR-122 was depleted rapidly from the hepatocytes (even after the 3rd or 4th 5 mg/kg dose). Comparing these data with Elmen et al. who reported that the nadir cholesterol level was reached approximately 3 weeks after dosing in the high-dose group, indicates that the effects on plasma cholesterol follows a slower rate in chimpanzees. Also the reduction in the viral titers was slow, and the nadir (week 12–14) occurred long after the estimated time where miR-122 was fully complexed. The reasons for this are not fully understood at this point, but speculations of an indirect mechanism of action in which miR-122 first has to be liberated/sequestered from the apparently strong Ago2/miR-122/HCV complex has been posted. After sequencing of the HCV RNA, no mutations in any of the miR-122 binding sites could be detected. Treating the chimpanzees with SPC3649 was well tolerated; for example, the liver toxicity marker alanine aminotransferase was reduced to normal levels, and the liver morphology was normalized, following the significant viral load reduction. SPC3649 is ongoing in clinical phase II as treatment for chronic infections with HCV.

4.2.4 LNA pharmacology in malignant tissues

Oncology is another area where pharmacology of LNAs has been demonstrated and where LNAs have been advanced to clinical trials. Today, three LNA-containing ASOs are being evaluated in the clinic, targeting different neoplastic indications. The mRNA targets of these three studies are: *Hif1-α* (Greenberger et al., 2008), *Survivin* (Hansen et al., 2008) (both against solid tumors), and the *AR* (Zhang et al., 2011) tested in patients with prostate cancer. In this section, we will focus on the pre-clinical models where these LNAs have been examined.

Hif1-α is a transcription factor, which plays a role in angiogenesis, metastasis, drug resistance, cell survival, and glucose metabolism. Greenberger et al. showed that EZN2968 was able to downregulate endogenous Hif1-α and vascular endothelial growth factor in the livers of normal mice (Greenberger et al., 2008). Dose response was observed with total doses of 50 mg/kg or 250 mg/kg in two different dose regiments. Administration of SPC2968 over 14 days with daily doses of either 3.6 mg/kg or 18 mg/kg led to the same significant response as 10 mg/kg or 50 mg/kg 5 times over 14 days. The effect could also be obtained after a single dose of 50 mg/kg, and the effect lasted up to 8 days after the administration. Finally, they found tumor reduction in nude mice implanted xenografts with DU145 prostate cancer cells following treatment with 10 doses of 50 mg/kg over 5 weeks.

Survivin is an apoptosis/mitosis control point in the cell and often activated during carcinogenesis. Hansen et al. showed that SPC3042/EZN3042, a 16-mer LNA against *Survivin*, was able to reduce tumor growth of the prostate cancer derived cell-line PC3 in xenotransplanted mice. The treatment was 20 mg/kg every 2 day for 19 days (Hansen et al., 2008). The most pronounced effect was seen in combination with the Taxol, which is a well-known inhibitor of the mitotic spindle formation. In the two preclinical models described here for Hif1-α and Survivin downregulation was not evaluated in the tumor tissue of the tested models.

One of the strongest target–phenotype correlations in oncology is the link between expression of the AR and progression of prostate cancer. EZN-4176 is a 16-mer LNA specific for targeting *AR* mRNA. Zhang et al. showed that EZN-4176, dosed at 60 mg/kg every 3 day for seven cycles, provided specific *AR* mRNA and protein down regulation (Zhang et al., 2011). The treatment was evaluated in nude mice in an androgen-dependent AR-positive CWR-22 tumor xenograft model. Growth of CWR-22 tumors, was inhibited to approximately 66% on day 27 similar to that observed with Casodex, an oral nonsteroidal antiandrogen used in the treatment of prostate cancer. The correlation between antitumor effect and AR status was established by showing inactivity of EZN-4176 in mice bearing AR-negative PC3 prostate tumors. When injected as a single dose of 60 mg/kg, the ASO showed a long residence time in tumors and sustained target downregulation. In addition, EZN-4176 (20 mg/kg) reduced tumor growth significantly in a castration-resistant prostate cancer (CRPC) animal model based on the C4-2b cell line in SCID mice. Luciferase reporter was used to visualize tumor growth *in vivo*. CRPC often metastasizes to the bone and EZN-4176 (40 mg/kg) was tested in the CRPC model with cells implanted intratibially, and showed significant and specific inhibition of the tumor.

5. LNA IN BIOTECHNOLOGY

5.1. LNA in primers and probes

5.1.1 LNA primers

Many biotechnological applications of ONs require sensitive and specific hybridization to complementary sequences, tasks which in certain cases cannot be fulfilled by unmodified ONs.

Thus, PCR can be improved by introducing LNA nts in the primers. In an early report, PCR amplification of the two human genes for ApoB and phenylanaline hydroxylase was investigated (Latorra, Arar, & Hurley, 2003). It was found that LNA primers offer several advantages over DNA primers, including higher maximal annealing temperature, improved performance with shorter primers and increasing the cost efficiency of PCR by lowering the amount of polymerase needed. However, not all LNA substitution-patterns improved PCR efficiency and it was generally found that too many LNA nts could be detrimental. Optimal results were observed when one to three LNA nts were placed centrally in the primer. In a later study, DNA primers were modified with LNA nts either toward the 5'- or the 3'-ends or evenly distributed over the length of the primer (Levin, Fiala, Samala, Kahn, & Peterson, 2006). It was found that primers substituted near the 5'-end generally performed better than the unmodified primer and LNA primers modified elsewhere. The superiority of LNA primers compared to DNA primers was confirmed by amplification experiments involving four routinely used short tandem repeat loci (Ballantyne, van Oorschot, & Mitchell, 2008). LNA primers had a broader tolerance to different reaction conditions and showed increased specificity. In addition, LNA primers resulted in decreased template requirements compared to unmodified DNA primers.

Real-time PCR amplification of low abundance mRNA can be problematic, especially if the sample is contaminated with DNA which gives rise to false positives. In the case of the *X-box binding protein 1* (*XBP1*) gene, primers may bind to the genomic DNA and produce an unwanted PCR product containing an intron. By binding LNA to the intron, amplification from DNA is blocked. The wanted amplification from the mRNA is not blocked, since it lacks the intron sequence (Hummelshoj, Ryder, Madsen, & Poulsen, 2005). This is an example how an LNA ON can be used to improve PCR without itself being directly involved in the amplification reactions.

In another study, detection by real-time PCR was investigated (Reynisson, Josefsen, Krause, & Hoorfar, 2006). *Taq*Man probes were compared to LNA and Scorpion probes. LNA probes outperformed the two other technologies by providing a higher fluorescence plateau, lower cycle threshold, and smaller standard deviation. The LNA probe was also robust since it was able to detect down to 10 template copies per reaction of *Salmonella* genomic DNA, on three different instruments. In a later study, LNA was confirmed as the preferred chemistry to be used in diagnostic PCR (Josefsen, Lofstrom, Sommer, & Hoorfar, 2009). LNA was superior due to its sensitivity toward single-base mismatches, ease of design and improved signal-to-noise ratio. More recently, a comparison between nine different real-time PCR chemistries was performed and it was found that none of the chemistries performed significantly better than the other (Buh Gasparic et al., 2010). However, the short LNA probes may offer advantages if high specificity is needed or if the target locus is short.

The discriminatory power of LNA ONs has also been used to detect single nt polymorphisms in genotyping experiments. In allele-specific reactions, primers must be extended only if they are fully complementary to their targets. Thus, LNA nts were introduced at the $3'$-position or one nt from the $3'$-end of the primer (Di Giusto & King, 2004). However, in this case only the LNA primer modified one nt from the $3'$-end showed a certain level of discrimination so as to inhibit the polymerase from extending when a single mismatch occurred at the extension site. However, when the assay was developed further by adding a proof-reading polymerase, the readout for the LNA primer modified one nt from the $3'$-end was clear. Correctly matched termini were extended, while no product was observed for incorrectly matched termini clearly enabling detection of the single nt polymorphism.

5.1.2 LNA probes

LNA probes were used in a qualitative fluorescence multiplex assay (Ugozzoli, Latorra, Puckett, Arar, & Hamby, 2004). Real-time PCR and $5'$-nuclease detection were combined to produce a four-color assay to detect single nt polymorphisms for factor V Leiden and prothrombin G20210A. LNA probes could be made shorter than the DNA counterparts, while still maintaining high melting temperature to increase specificity. The LNA probes clearly outperformed DNA probes to give signal only when a completely matched target was present.

In general, the unique hybridization capability observed for LNA ONs make them useful for detection of noncoding RNAs (Stenvang, Silahtaroglu, Lindow, Elmen, & Kauppinen, 2008). LNA probes can be designed to have equal melting temperatures for their respective miRNA target. In this way, a microarray was developed in which LNA ONs function as capture probes for miRNA (Castoldi, Benes, Hentze, & Muckenthaler, 2007). The microarray was robust and able to discriminate between closely related miRNAs.

With the aim of improving the sensitivity and resolution of fluorescence *in situ* hybridization, LNA probes have also been produced (Silahtaroglu, Pfundheller, Koshkin, Tommerup, & Kauppinen, 2004). These probes turned out to work excellently by combining high binding affinity with short hybridization times. The sensitivity and performance of LNA probes were fully comparable to PNA probes. Also, to determine spatial and temporal miRNA accumulation in different tissues, *in situ* hybridization methods with LNA probes can be used (Kloosterman, Wienholds, de Bruijn, Kauppinen, & Plasterk, 2006). Further improvement has been achieved by combining LNA monomers with $2'$-O-methyl-RNA monomers to give probes with superior efficiency, for example, in detection of miR-146a via *in situ* hybridization (Aronica et al., 2010).

5.2. LNA-containing molecular beacons

Molecular beacons are stem-loop structures that can be used to detect DNA. A fluorophore and a quencher are attached to either end of the beacon. Without target, the fluorescence is quenched but the structure unfolds in the presence of the target and the fluorescence is increased. Limitations exist when visualizing expression in living cells due to degradation and binding to proteins, both of which give rise to false positive signals. To remedy this, molecular beacons with LNA nts in the stem of the beacon have been constructed and compared to the equivalent unmodified beacon (Wang, Yang, Medley, Benner, & Tan, 2005). After hybridization, the tight binding induced by LNA monomers resulted in signal enhancement at both 25 and 95 °C. The LNA molecular beacon was also able to better discriminate between fully matched and a single mismatched targets, as compared to the unmodified beacon. The kinetics were slow due to the high energy barrier of opening the stem containing LNA nts, however, a shorter stem resulted in faster kinetics without loss of specificity. The presence of LNA nts further provided significant protection against DNaseI, whereas the unmodified molecular beacon was quickly degraded. A false positive signal was observed when the unmodified beacon, but not when the LNA beacon, was exposed to single-stranded DNA-binding protein. When gene expression was measured in an intracellular environment, the two molecular beacons gave similar responses to the target. However, in the absence of the target, the signal for the LNA beacon was very low while the unmodified beacon gave a significant false positive signal.

Another publication dealt with molecular beacons immobilized on solid surfaces (Martinez et al., 2009). The fluorescence enhancement of molecular beacons is not as high when immobilized to a surface as it is in solution. For this reason an LNA, molecular beacon was produced, as the rigidity of the LNA pairing was expected to prevent interactions with the surface. Indeed, it was found that the background for the LNA molecular beacon was lower than for the unmodified beacon, and given the right distance to the surface, upon target binding a 25-fold fluorescence enhancement was observed with the LNA beacon. For the LNA beacon, the background was constant over a temperature range of 4–50 °C, which was not the case for the more thermally unstable unmodified beacon. Although the LNA beacon displayed slower kinetics than the unmodified ON, signal was still observed within a few minutes. The LNA beacon was also able to detect lower levels of the target and displayed a higher specificity than the unmodified molecular beacon. The fluorescence enhancement dropped slightly, but not dramatically, when the target was present in complex matrices such as fetal bovine serum or a cell lysate.

5.3. LNA modification in aptamers

Nucleic acid aptamers are short single-stranded molecules originally made of DNA or RNA that fold into three-dimensional structures, which enable them to bind a certain target with high affinity and specificity (Syed & Pervaiz, 2010). In this way, aptamers are reminiscent of antibodies in their mode of action; however, they offer many advantages compared to antibodies. Aptamers have been shown to be nonimmunogenic and reversal of therapeutic effect can be achieved in certain cases; they are robustly manufactured by standard phosphoramidite chemistry and they can be modified in a variety of ways to improve PKs (Syed & Pervaiz, 2010). Aptamers have been developed against a wide array of targets including ions, small molecules, proteins, viruses and cells, and typically bind to the target with a dissociation constant in the nanomolar range. Aptamers are typically developed by an iterative *in vitro* selection method known as Systematic Evolution of Ligands by EXponential enrichment (SELEX). In SELEX, a large library of nucleic acid sequences is generated by standard phoshoramidite chemistry. The library is exposed to the target of interest and nonbinding species are washed away. After recovery of the binding species from the target, PCR is performed to amplify the species in the binding pool. Once the aptamer strand has been regenerated, another round of SELEX can be performed. By increasing the selection pressure, the aptamer pool will converge into relatively few sequences with high affinity for the target.

Aptamers have many potential applications relating to diagnostics, purification, target validation, and therapeutics (Proske, Blank, Buhmann, & Resch, 2005). Still, they must be modified for general therapeutic applications since nucleases and renal filtration quickly clear an aptamer from the blood stream. Mainly due to the nuclease resistance offered by LNA, a number of aptamers have been modifed with LNA nts. An early example of an LNA-modified aptamer was reported in 2004 (Darfeuille, Hansen, Ørum,

Di Primo, & Toulme, 2004), when LNA nts were introduced into the RNA aptamer R06. This aptamer binds to the *trans*-activating responsive (TAR) RNA element of HIV-1, in an interaction which includes base pairing. This loop–loop interaction interferes with binding of the *trans*-activator protein to TAR, ultimately leading to inhibition of virus replication. LNA/DNA chimeras of this aptamer were generally well protected from nucleases, and one chimera behaved as the original RNA aptamer itself. In a later study, six positions in the loop were systematically modified with either LNA or 2′OMe RNA nts (Di Primo et al., 2007). It was found that three of the combinations in which one or two LNA nts are located on the 3′ side of the loop provided aptamers with affinities one order of magnitude better the original RNA aptamer. In addition, TAR-dependent luciferase expression was inhibited in a cell.

LNA nts have also been introduced into the TTA1 aptamer, which binds Tenascin-C (Schmidt et al., 2004). This aptamer folds into a three-way junction creating a structure with three stem regions. Binding affinity is completely lost when stems II and III contain LNA nts; however, LNA monomers are tolerated in stem I. In addition, this LNA aptamer was significantly protected from nuclease degradation. Compared to the unmodified aptamer, the LNA-containing version remained longer in the blood and displayed higher uptake in tumors in nude mice.

Modifying the thrombin-binding aptamer (TBA) with LNA nts has been less successful (Virno et al., 2007). TBA is a DNA aptamer that folds into a G-quadruplex. Substituting the G in the 3′-end of the aptamer resulted in reduced anticoagulant activity compared to the unmodified aptamer. In a further study, a single LNA nt was introduced into the middle or toward the 5′-end of the TBA, but binding affinity was not improved (Bonifacio, Church, & Jarstfer, 2008). It was shown that thermal stability did not translate into thrombin-binding affinity, indicating that specific interactions between the nts in TBA and thrombin provide binding rather than thermal stability of the G-quadruplex itself. However, the G2 position of TBA could harbor an LNA nt monomer without significant loss of activity.

Sgc8c is an aptamer developed against a human T-cell acute lymphoblastic leukemia cell line, and is assumed to fold into a stem-loop structure (Shangguan, Tang, Mallikaratchy, Xiao, & Tan, 2007). When replacing all nts in the sequence or all nts in the stem with LNA monomers, binding affinity is completely lost. However, when only three base pairs were replaced, the binding affinity was close to that of unmodified sgc8c, and stability in cell culture medium was increased. These LNA modifications protected well against exonucleases; however, endonucleases were still able to degrade the aptamer.

In another report, a DNA avidin-binding aptamer was truncated to 21 nts and the influence of single or double incorporations of LNA nts at different positions in the sequence was investigated (Hernandez, Kalra, Wengel, & Vester, 2009). In general, LNA was tolerated well and most modified sequences gave similar responses as the unmodified DNA aptamer. The binding affinity of ONs modified at position G2 was further investigated. An LNA nt at position G2 resulted in almost an order of magnitude lower dissociation constant than the unmodified aptamer. A $2'$-OMe nt substitution did not give the same effect, indicating that the unique structure of LNA is important.

A stem-loop aptamer for the B-cell receptor called TD05 was recently truncated slightly to decrease the number of nts in the stem (Mallikaratchy et al., 2011). The influence of LNA nts in the stem was tested, by substituting four pyrimidines or four purines with LNA nts. The pyrimidines could be replaced and an increased binding was even observed, in contrast to what was found when the purines opposite to the pyrimidines were modified. This indicates that the pyrimidines of the stem stabilize the structure but are not directly involved in target binding. As expected, LNA nt substitutions were found to increase conformational stability and nuclease resistance. Recently, the RNA aptamer for ricin was also modified with LNA nts (Forster et al., 2012). Aptamer RA80.1.d1 with a stem-loop structure has a three nt bulge on the long stem. The stem was shortened and at least 14 LNA nts were introduced into the stem. When LNA nts occurred in the bulge region, binding to ricin was completely abolished. The binding affinity rose as the LNA nts were placed further away from the bulge. However, the binding response was not as high as for the original RNA aptamer with the long stem. The best LNA aptamer showed increased resistance to ribonucleases V1, A, T1, and R compared to the naked RNA aptamer.

In general, LNA nts provide protection from nucleases for the aptamer in which they are incorporated. Modifying the stem of stem-loop structures is usually well tolerated though binding affinity may be lost in certain cases. Generally, modification of known SELEX-evolved aptamers with LNA nts is a trial-and-error process. Research is currently being performed to include LNA nts as part of the selection procedure. For this purpose, polymerases that can work with LNA nts are needed in order to amplify the selection pools by PCR (see next section).

5.4. LNA in enzymatic reactions

5.4.1 Incorporation of LNA by DNA polymerases

In a recent report, mutated polymerases were developed that were able to work with unnatural nucleic acid backbones, the so-called xeno-nucleic acids (XNAs) (Pinheiro et al., 2012). Using a technique known as compartmentalized self-tagging, polymerases with novel substrate specificities were evolved. Mutations were introduced into a gene encoding for TgoT, which is a variant of the replicative polymerase of *Thermococcus gorgonarius*. The mutated variants were then selected for their ability to work with LNA and other XNAs. Two particular polymerase variants of TgoT polymerase were particularly interesting when it comes to LNA replication. The polymerase variants called PolC7 and RT521K contained 10 and 6 amino acid substitutions of the original TgoT polymerase, respectively. PolC7 was able to synthesize LNA ONs from a DNA template. Conversely, RT521K could reverse transcribe LNA, meaning that LNA ON templates could be translated back to DNA. This shows that LNA, like DNA and RNA, has the potential to function as genetic material capable of heredity and evolution.

Although natural polymerases obviously have not been selected to be compatible with LNA nts, a number of commercially available polymerases are also able to work with LNA. The incorporation and reading of LNA nts can be followed by extension reactions on labeled primers. In two early reports, several commercially available polymerases were screened for their ability to use LNA ATP and LNA TTP in primer extension reactions (Veedu, Vester, & Wengel, 2007a, 2007b). Most polymerases performed unsatisfactory. Phusion DNA polymerase, however, was able to extend the primer to full length in certain cases. Incorporation of LNA-A and LNA-T was accomplished when LNA nt incorporation was followed by one or more DNA nts. In particular, the Phusion DNA polymerase had problems with incorporating consecutive LNA-T nts. This was also confirmed in a later study (Kuwahara et al., 2008). Another investigation revealed that Phusion DNA polymerase could incorporate LNA nts when the templating nt was not only a DNA but also an LNA nt (Veedu, Vester, & Wengel, 2008). PCR amplification requiring both incorporation and reading of LNA-A nts was also achieved. In addition, the $9°N_m$ DNA polymerase performed equally well as Phusion DNA polymerase at incorporating LNA-A opposite both DNA and LNA templating thymine bases. However, degradation of DNA products by the inherent nucleolytic activity of this polymerase may limit the general applicability of the $9°N_m$ DNA polymerase (Veedu et al., 2008). Several other polymerases were also tested for their compatibility with LNA and other locked nts (Kuwahara et al., 2008). Templates containing LNA nts were designed to determine how far apart modified nts need to be in order to allow full-length extension. One template contained consecutive LNA nts, while other templates had one, two, or three natural nts separating the LNA nts. It was found that none of the polymerases screened were able to work with several consecutive LNA nts in the template; however, KOD Dash, KOD(exo-), and Vent (exo-) extended the primer to full length when LNA and natural nts were interdispersed in the template. KOD Dash and Phusion DNA polymerases were further tested for their ability to extend the primer with several LNA nts using a DNA template. However, none of the two polymerases was able to extend the primer to full length.

The most efficient incorporation of LNA nts to date has been reported for KOD DNA polymerase (Veedu, Vester, & Wengel, 2009). Like Phusion and $9°N_m$ DNA polymerases, KOD can incorporate LNA-A and LNA-T across from both DNA and LNA template nts. In addition, KOD can use LNA 5′-methyl-CTP as a substrate and incorporation of even eight consecutive LNA-^{5Me}C incorporations was unproblematic. Still, in an extension experiment using a mix of GTP and three LNA NTPs, the polymerase stalled prematurely. Recently, incorporation of LNA-G nts was however achieved by KOD DNA polymerase (Veedu, Vester, & Wengel, 2010). Amazingly, a primer could be extended to full length using only LNA NTPs on a DNA template leading to incorporation of 21 consecutive LNA nts. In addition, KOD DNA polymerase has also a potential for LNA amplification as a PCR incorporating LNA-A nts was successful (Veedu et al., 2009).

5.4.2 Incorporation of LNA by RNA polymerases

Incorporation of LNA nts is not limited to DNA polymerases (Kore, Hodeib, & Hu, 2008; Veedu et al., 2008). T7 RNA polymerase was able to produce full-length RNA transcripts containing up to eight consecutive incorporations of LNA-A nts. Moreover, T7 RNA polymerase could incorporate both LNA nts as well as ribonucleotides across from LNA nts in the template. This demonstrates that LNA can not only be replicated but also be transcribed by polymerases. In addition, a recent report found that Super-Script III Reverse Transcriptase works well when it comes to reverse transcribing LNA (Crouzier et al., 2012). The polymerase could incorporate LNA nts into a DNA strand using an RNA template. Conversely, the polymerase was also able to use an RNA template containing LNA nts to incorporate both LNA nts and deoxynucleotides across from the LNAs. The capacity of certain polymerases to both read and incorporate LNA nts during PCR-like conditions is highly important in the biotechnological perspective. These manipulations are vital for LNA-aptamer evolution, as they are prerequisites for including LNA nts in the SELEX process.

6. CONCLUDING REMARKS

From the description of the numerous fields, where LNA currently is under study, it is apparent that this chemistry has already had a profound influence in many areas, ranging from biotechnology to drug development. It should be mentioned that while additional chemical modifications of the LNA backbone have been generated, many of these have not been extensively tested. Since the chemical universe is wide, a huge potential for optimization of the LNA backbone exists *en route* to optimize LNA-based constructs for specific applications.

The research around LNA ONs as antigene compounds is still in its infancy, in contrast to the developments in the transcript antisense field. The pharmacology of LNA ASOs has been extensively explored over the past 12 years. LNAs have been tested in many different animal species exhibiting pharmacology against a variety of targets expressed in a multitude of tissues, and seven LNAs have been tested in clinical trials. The most advanced of these is SPC3649 (Miravirsen) that has completed a phase IIa study (Janssen et al., 2013). This antisense LNA is a potent inhibitor of microRNA-122, a crucial host factor for HCV. This is the first microRNA inhibitor tested in man and interestingly, exhibited marked reductions in HCV titers in infected individuals. The data obtained from this study validate not only this specific compound clinically but also importantly validates LNA ONs as a drug platform in a broader sense.

Within other biotechnological fields, the use of LNA has also experienced rapid developments. For instance, the use of LNA aptamers emerges as a highly interesting technology, both for research purposes and for medical approaches. Currently, aptamers are rather expensive to identify and the algorithms for predicting structural changes upon substitution of, for example, DNA or RNA bases for LNA are of insufficient quality. Improvements along these lines would profoundly change the LNA-aptamer field and potentially allow for numerous applications. Furthermore, the possibility of enzymatic incorporation of LNA would enable SELEX-based evolution of LNA-containing aptamers. Although much of the LNA universe has yet to be explored, we have already seen considerable progress and remarkable achievements in the applications of LNA in both research and therapeutic fields. The future bodes well for this unique bicyclic nucleoside chemistry.

ACKNOWLEDGMENTS

The authors would like to thank the EuroNanoMed Joint Transnational project Nanosplice, the Swedish Research Council, the CHDI foundation, Söderbergs foundation, the Lundbeck Foundation, and the ERC for financial support. J. K. was supported by Aarhus University and J. B. B. by The Lundbeck Center for Nanomedicine.

Part of this review is based on the works over many years from many highly skilled scientists and technicians at Santaris Pharma A/S. Without their devoted effort these parts would not have been possible, and we would like to thank them for their contribution. The authors would also like to thank Principal Scientist Henrik F. Hansen for commenting to these parts of the chapter.

Conflict of interest statement. Some of the work presented here is made at Santaris Pharma, and the authors Bo Rode Hansen, Robert Persson, and Troels Koch are full-time employees of Santaris Pharma.

REFERENCES

Aartsma-Rus, A. (2012). Overview on aon design. *Methods in Molecular Biology, 867,* 117–129.
Aartsma-Rus, A., Kaman, W. E., Bremmer-Bout, M., Janson, A. A., den Dunnen, J. T., van Ommen, G. J., et al. (2004). Comparative analysis of antisense oligonucleotide analogs for targeted dmd exon 46 skipping in muscle cells. *Gene Therapy, 11,* 1391–1398.
Afonin, K. A., Grabow, W. W., Walker, F. M., Bindewald, E., Dobrovolskaia, M. A., Shapiro, B. A., et al. (2011). Design and self-assembly of siRNA-functionalized RNA nanoparticles for use in automated nanomedicine. *Nature Protocols, 6,* 2022–2034.
Akhtar, S., Basu, S., Wickstrom, E., & Juliano, R. L. (1991). Interactions of antisense DNA oligonucleotide analogs with phospholipid membranes (liposomes). *Nucleic Acids Research, 19,* 5551–5559.
Alam, M. R., Majumdar, A., Thazhathveetil, A. K., Liu, S. T., Liu, J. L., Puri, N., et al. (2007). Extensive sugar modification improves triple helix forming oligonucleotide activity in vitro but reduces activity in vivo. *Biochemistry, 46,* 10222–10233.
Alam, M. R., Ming, X., Dixit, V., Fisher, M., Chen, X., & Juliano, R. L. (2010). The biological effect of an antisense oligonucleotide depends on its route of endocytosis and trafficking. *Oligonucleotides, 20,* 103–109.
Alexopoulou, L., Holt, A. C., Medzhitov, R., & Flavell, R. A. (2001). Recognition of double-stranded RNA and activation of nf-kappab by toll-like receptor 3. *Nature, 413,* 732–738.
Alvira, M., & Eritja, R. (2010). Triplex-stabilizing properties of parallel clamps carrying LNA derivatives at the hoogsteen strand. *Chemistry & Biodiversity, 7,* 376–382.
Aronica, E., Fluiter, K., Iyer, A., Zurolo, E., Vreijling, J., van Vliet, E. A., et al. (2010). Expression pattern of mir-146a, an inflammation-associated microRNA, in experimental and human temporal lobe epilepsy. *The European Journal of Neuroscience, 31,* 1100–1107.

Astakhova, I. V., Korshun, V. A., Jahn, K., Kjems, J., & Wengel, J. (2008). Perylene attached to 2′-amino-LNA: Synthesis, incorporation into oligonucleotides, and remarkable fluorescence properties in vitro and in cell culture. *Bioconjugate Chemistry, 19,* 1995–2007.

Ballantyne, K. N., van Oorschot, R. A. H., & Mitchell, R. J. (2008). Locked nucleic acids: Increased trace DNA amplification success with improved primers. *Forensic Science International: Genetics Supplement Series, 1,* 4–6.

Beane, R., Gabillet, S., Montaillier, C., Arar, K., & Corey, D. R. (2008). Recognition of chromosomal DNA inside cells by locked nucleic acids (dagger). *Biochemistry, 47,* 13147–13149.

Beane, R. L., Ram, R., Gabillet, S., Arar, K., Monia, B. P., & Corey, D. R. (2007). Inhibiting gene expression with locked nucleic acids (Lnas) that target chromosomal DNA. *Biochemistry, 46,* 7572–7580.

Bentin, T., & Nielsen, P. E. (1996). Enhanced peptide nucleic acid binding to supercoiled DNA: Possible implications for DNA "breathing" dynamics. *Biochemistry, 35,* 8863–8869.

Bertrand, J. R., Pottier, M., Vekris, A., Opolon, P., Maksimenko, A., & Malvy, C. (2002). Comparison of antisense oligonucleotides and siRNAs in cell culture and in vivo. *Biochemical and Biophysical Research Communications, 296,* 1000–1004.

Bestas, B., Moreno, P. M., Blomberg, E. M., Berglöf, A., Gustafsson, M. O., Mohammad, D. K., et al. (2012). RNA therapeutics in xla—A novel strategy to repair btk. In: *15th Biennial meeting of the European Society for immunodeficiencies florence.*

Bonifacio, L., Church, F. C., & Jarstfer, M. B. (2008). Effect of locked-nucleic acid on a biologically active g-quadruplex. A structure-activity relationship of the thrombin aptamer. *International Journal of Molecular Science, 9,* 422–433.

Braasch, D. A., Jensen, S., Liu, Y., Kaur, K., Arar, K., White, M. A., et al. (2003). RNA interference in mammalian cells by chemically-modified RNA. *Biochemistry, 42,* 7967–7975.

Braasch, D. A., Liu, Y., & Corey, D. R. (2002). Antisense inhibition of gene expression in cells by oligonucleotides incorporating locked nucleic acids: Effect of mRNA target sequence and chimera design. *Nucleic Acids Research, 30,* 5160–5167.

Bramsen, J. B., & Kjems, J. (2012). Development of therapeutic-grade small interfering RNAs by chemical engineering. *Frontiers in Genetics, 3,* 154.

Bramsen, J. B., Laursen, M. B., Damgaard, C. K., Lena, S. W., Babu, B. R., Wengel, J., et al. (2007). Improved silencing properties using small internally segmented interfering RNAs. *Nucleic Acids Research, 35,* 5886–5897.

Bramsen, J. B., Laursen, M. B., Nielsen, A. F., Hansen, T. B., Bus, C., Langkjær, N., et al. (2009). A large-scale chemical modification screen identifies design rules to generate siRNAs with high activity, high stability and low toxicity. *Nucleic Acids Research, 37,* 2867–2881.

Brunet, E., Alberti, P., Perrouault, L., Babu, R., Wengel, J., & Giovannangeli, C. (2005). Exploring cellular activity of locked nucleic acid-modified triplex-forming oligonucleotides and defining its molecular basis. *The Journal of Biological Chemistry, 280,* 20076–20085.

Brunet, E., Corgnali, M., Cannata, F., Perrouault, L., & Giovannangeli, C. (2006). Targeting chromosomal sites with locked nucleic acid-modified triplex-forming oligonucleotides: Study of efficiency dependence on DNA nuclear environment. *Nucleic Acids Research, 34,* 4546–4553.

Brunet, E., Corgnali, M., Perrouault, L., Roig, V., Asseline, U., Sørensen, M. D., et al. (2005). Intercalator conjugates of pyrimidine locked nucleic acid-modified triplex-forming oligonucleotides: Improving DNA binding properties and reaching cellular activities. *Nucleic Acids Research, 33,* 4223–4234.

Bruylants, G., Boccongelli, M., Snoussi, K., & Bartik, K. (2009). Comparison of the thermodynamics and base-pair dynamics of a full LNA:DNA duplex and of the isosequential DNA:DNA duplex. *Biochemistry*, *48*, 8473–8482.

Bryld, T., & Lomholt, C. (2007). Attatchment of cholesterol to amino-LNA: Synthesis and hybridization properties. *Nucleosides, Nucleotides & Nucleic Acids*, *26*, 1645–1647.

Buh Gasparic, M., Tengs, T., La Paz, J. L., Holst-Jensen, A., Pla, M., Esteve, T., et al. (2010). Comparison of nine different real-time pcr chemistries for qualitative and quantitative applications in gmo detection. *Analytical and Bioanalytical Chemistry*, *396*, 2023–2029.

Burnett, J. C., Rossi, J. J., & Tiemann, K. (2011). Current progress of siRNA/shRNA therapeutics in clinical trials. *Biotechnology Journal*, *6*, 1130–1146.

Castoldi, M., Benes, V., Hentze, M. W., & Muckenthaler, M. U. (2007). Michip: A microarray platform for expression profiling of microRNAs based on locked nucleic acid (LNA) oligonucleotide capture probes. *Methods*, *43*, 146–152.

Chiu, Y. L., & Rana, T. M. (2002). RNAi in human cells: Basic structural and functional features of small interfering RNA. *Molecular Cell*, *10*, 549–561.

Chiu, Y. L., & Rana, T. M. (2003). SiRNA function in RNAi: A chemical modification analysis. *RNA*, *9*, 1034–1048.

Cirak, S., Arechavala-Gomeza, V., Guglieri, M., Feng, L., Torelli, S., Anthony, K., et al. (2011). Exon skipping and dystrophin restoration in patients with duchenne muscular dystrophy after systemic phosphorodiamidate morpholino oligomer treatment: An open-label, phase 2, dose-escalation study. *Lancet*, *378*, 595–605.

Collingwood, M. A., Rose, S. D., Huang, L., Hillier, C., Amarzguioui, M., Wiiger, M. T., et al. (2008). Chemical modification patterns compatible with high potency dicer-substrate small interfering RNAs. *Oligonucleotides*, *18*, 187–200.

Corsten, M. F., Miranda, R., Kasmieh, R., Krichevsky, A. M., Weissleder, R., & Shah, K. (2007). MicroRNA-21 knockdown disrupts glioma growth in vivo and displays synergistic cytotoxicity with neural precursor cell delivered s-trail in human gliomas. *Cancer Research*, *67*, 8994–9000.

Crinelli, R., Bianchi, M., Gentilini, L., & Magnani, M. (2002). Design and characterization of decoy oligonucleotides containing locked nucleic acids. *Nucleic Acids Research*, *30*, 2435–2443.

Crouzier, L., Dubois, C., Edwards, S. L., Lauridsen, L. H., Wengel, J., & Veedu, R. N. (2012). Efficient reverse transcription using locked nucleic acid nucleotides towards the evolution of nuclease resistant RNA aptamers. *PLoS One*, *7*, e35990.

Darfeuille, F., Hansen, J. B., Ørum, H., Di Primo, C., & Toulme, J. J. (2004). LNA/DNA chimeric oligomers mimic RNA aptamers targeted to the tar RNA element of hiv-1. *Nucleic Acids Research*, *32*, 3101–3107.

Davis, S., Lollo, B., Freier, S., & Esau, C. (2006). Improved targeting of miRNA with antisense oligonucleotides. *Nucleic Acids Research*, *34*, 2294–2304.

Demidov, V. V., Protozanova, E., Izvolsky, K. I., Price, C., Nielsen, P. E., & Frank-Kamenetskii, M. D. (2002). Kinetics and mechanism of the DNA double helix invasion by pseudocomplementary peptide nucleic acids. *Proceedings of the National Academy of Sciences of the United States of America*, *99*, 5953–5958.

Di Giusto, D. A., & King, G. C. (2004). Strong positional preference in the interaction of LNA oligonucleotides with DNA polymerase and proofreading exonuclease activities: Implications for genotyping assays. *Nucleic Acids Research*, *32*, e32.

Di Primo, C., Rudloff, I., Reigadas, S., Arzumanov, A. A., Gait, M. J., & Toulme, J. J. (2007). Systematic screening of LNA/2′-o-methyl chimeric derivatives of a tar RNA aptamer. *FEBS Letters*, *581*, 771–774.

Diebold, S. S., Kaisho, T., Hemmi, H., Akira, S., & Reis e Sousa, C. (2004). Innate antiviral responses by means of TLR7-mediated recognition of single-stranded RNA. *Science*, *303*, 1529–1531.

Diebold, S. S., Massacrier, C., Akira, S., Paturel, C., Morel, Y., & Reis e Sousa, C. (2006). Nucleic acid agonists for toll-like receptor 7 are defined by the presence of uridine ribonucleotides. *European Journal of Immunology, 36*, 3256–3267.

Dominski, Z., & Kole, R. (1993). Restoration of correct splicing in thalassemic pre-mRNA by antisense oligonucleotides. *Proceedings of the National Academy of Sciences of the United States of America, 90*, 8673–8677.

Dowler, T., Bergeron, D., Tedeschi, A. L., Paquet, L., Ferrari, N., & Damha, M. J. (2006). Improvements in siRNA properties mediated by 2′-deoxy-2′-fluoro-beta-d-arabinonucleic acid (fana). *Nucleic Acids Research, 34*, 1669–1675.

Dutkiewicz, M., Grunert, H. P., Zeichhardt, H., Lena, S. W., Wengel, J., & Kurreck, J. (2008). Design of LNA-modified siRNAs against the highly structured 5′ utr of coxsackievirus b3. *FEBS Letters, 582*, 3061–3066.

Eberle, F., Giessler, K., Deck, C., Heeg, K., Peter, M., Richert, C., et al. (2008). Modifications in small interfering RNA that separate immunostimulation from RNA interference. *Journal of Immunology (Baltimore, Md.: 1950), 180*, 3229–3237.

Eckstein, F. (2007). The versatility of oligonucleotides as potential therapeutics. *Expert Opinion on Biological Therapy, 7*, 1021–1034.

Egholm, M., Christensen, L., Dueholm, K. L., Buchardt, O., Coull, J., & Nielsen, P. E. (1995). Efficient ph-independent sequence-specific DNA binding by pseudoisocytosine-containing bis-pna. *Nucleic Acids Research, 23*, 217–222.

El Andaloussi, S. A., Hammond, S. M., Mager, I., & Wood, M. J. (2012). Use of cell-penetrating-peptides in oligonucleotide splice switching therapy. *Current Gene Therapy, 12*, 161–178.

Elmen, J., Lindow, M., Schutz, S., Lawrence, M., Petri, A., Obad, S., et al. (2008). LNA-mediated microRNA silencing in non-human primates. *Nature, 452*, 896–899.

Elmen, J., Lindow, M., Silahtaroglu, A., Bak, M., Christensen, M., Lind-Thomsen, A., et al. (2008). Antagonism of microRNA-122 in mice by systemically administered LNA-antimir leads to up-regulation of a large set of predicted target mRNAs in the liver. *Nucleic Acids Research, 36*, 1153–1162.

Elmén, J., Thonberg, H., Ljungberg, K., Frieden, M., Westergaard, M., Xu, Y., et al. (2005). Locked nucleic acid (LNA) mediated improvements in siRNA stability and functionality. *Nucleic Acids Research, 33*, 439–447.

Emmrich, S., Wang, W., John, K., Li, W., & Putzer, B. M. (2009). Antisense gapmers selectively suppress individual oncogenic p73 splice isoforms and inhibit tumor growth in vivo. *Molecular Cancer, 8*, 61.

Esau, C., Davis, S., Murray, S. F., Yu, X. X., Pandey, S. K., Pear, M., et al. (2006). Mir-122 regulation of lipid metabolism revealed by in vivo antisense targeting. *Cell Metabolism, 3*, 87–98.

Fedorov, Y., Anderson, E. M., Birmingham, A., Reynolds, A., Karpilow, J., Robinson, K., et al. (2006). Off-target effects by siRNA can induce toxic phenotype. *RNA, 12*, 1188–1196.

Fluiter, K., Frieden, M., Vreijling, J., Koch, T., & Baas, F. (2005). Evaluation of LNA-modified DNAzymes targeting a single nucleotide polymorphism in the large subunit of RNA polymerase ii. *Oligonucleotides, 15*, 246–254.

Fluiter, K., Frieden, M., Vreijling, J., Rosenbohm, C., De Wissel, M. B., Christensen, S. M., et al. (2005). On the in vitro and in vivo properties of four locked nucleic acid nucleotides incorporated into an anti-h-ras antisense oligonucleotide. *ChemBioChem, 6*, 1104–1109.

Fluiter, K., Housman, D., Ten Asbroek, A. L., & Baas, F. (2003). Killing cancer by targeting genes that cancer cells have lost: Allele-specific inhibition, a novel approach to the treatment of genetic disorders. *Cellular and Molecular Life Sciences, 60*, 834–843.

Fluiter, K., ten Asbroek, A. L., de Wissel, M. B., Jakobs, M. E., Wissenbach, M., Olsson, H., et al. (2003). In vivo tumor growth inhibition and biodistribution studies of locked nucleic acid (LNA) antisense oligonucleotides. *Nucleic Acids Research, 31*, 953–962.

Forsbach, A., Nemorin, J. G., Montino, C., Muller, C., Samulowitz, U., Vicari, A. P., et al. (2008). Identification of RNA sequence motifs stimulating sequence-specific TLR8-dependent immune responses. *Journal of Immunology (Baltimore, Md.: 1950), 180*, 3729–3738.

Forster, C., Zydek, M., Rothkegel, M., Wu, Z., Gallin, C., Gessner, R., et al. (2012). Properties of an LNA-modified ricin RNA aptamer. *Biochemical and Biophysical Research Communications, 419*, 60–65.

Frank-Kamenetskii, M. (1987). DNA chemistry. How the double helix breathes. *Nature, 328*, 17–18.

Frank-Kamenetskii, M. D., & Mirkin, S. M. (1995). Triplex DNA structures. *Annual Review of Biochemistry, 64*, 65–95.

Frieden, M., Christensen, S. M., Mikkelsen, N. D., Rosenbohm, C., Thrue, C. A., Westergaard, M., et al. (2003). Expanding the design horizon of antisense oligonucleotides with alpha-l-LNA. *Nucleic Acids Research, 31*, 6365–6372.

Gagnon, K. T., Watts, J. K., Pendergraff, H. M., Montaillier, C., Thai, D., Potier, P., et al. (2011). Antisense and antigene inhibition of gene expression by cell-permeable oligonucleotide-oligospermine conjugates. *Journal of the American Chemical Society, 133*, 8404–8407.

Gao, S., Dagnaes-Hansen, F., Nielsen, E. J., Wengel, J., Besenbacher, F., Howard, K. A., et al. (2009). The effect of chemical modification and nanoparticle formulation on stability and biodistribution of siRNA in mice. *Molecular Therapy, 17*, 1225–1233.

Ge, R., Heinonen, J. E., Svahn, M. G., Mohamed, A. J., Lundin, K. E., & Smith, C. I. (2007). Zorro locked nucleic acid induces sequence-specific gene silencing. *The FASEB Journal, 21*, 1902–1914.

Ge, R., Svahn, M. G., Simonson, O. E., Mohamed, A. J., Lundin, K. E., & Smith, C. I. (2008). Sequence-specific inhibition of RNA polymerase iii-dependent transcription using zorro locked nucleic acid (LNA). *The Journal of Gene Medicine, 10*, 101–109.

Geary, R. S., Wancewicz, E., Matson, J., Pearce, M., Siwkowski, A., Swayze, E., et al. (2009). Effect of dose and plasma concentration on liver uptake and pharmacologic activity of a 2′-methoxyethyl modified chimeric antisense oligonucleotide targeting pten. *Biochemical Pharmacology, 78*, 284–291.

Glud, S. Z., Bramsen, J. B., Dagnaes-Hansen, F., Wengel, J., Howard, K. A., Nyengaard, J. R., et al. (2009). Naked siLNA-mediated gene silencing of lung bronchoepithelium egfp expression after intravenous administration. *Oligonucleotides, 19*, 163–168.

Goemans, N. M., Tulinius, M., van den Akker, J. T., Burm, B. E., Ekhart, P. F., Heuvelmans, N., et al. (2011). Systemic administration of pro051 in duchenne's muscular dystrophy. *The New England Journal of Medicine, 364*, 1513–1522.

Goodchild, A., Nopper, N., King, A., Doan, T., Tanudji, M., Arndt, G. M., et al. (2009). Sequence determinants of innate immune activation by short interfering RNAs. *BMC Immunology, 10*, 40.

Graham, M. J., Lemonidis, K. M., Whipple, C. P., Subramaniam, A., Monia, B. P., Crooke, S. T., et al. (2007). Antisense inhibition of proprotein convertase subtilisin/kexin type 9 reduces serum ldl in hyperlipidemic mice. *Journal of Lipid Research, 48*, 763–767.

Graziewicz, M. A., Tarrant, T. K., Buckley, B., Roberts, J., Fulton, L., Hansen, H., et al. (2008). An endogenous tnf-alpha antagonist induced by splice-switching oligonucleotides reduces inflammation in hepatitis and arthritis mouse models. *Molecular Therapy, 16*, 1316–1322.

Greenberger, L. M., Horak, I. D., Filpula, D., Sapra, P., Westergaard, M., Frydenlund, H. F., et al. (2008). A RNA antagonist of hypoxia-inducible factor-1alpha, ezn-2968, inhibits tumor cell growth. *Molecular Cancer Therapeutics, 7*, 3598–3608.

Gueron, M., Kochoyan, M., & Leroy, J. L. (1987). A single mode of DNA base-pair opening drives imino proton exchange. *Nature, 328*, 89–92.

Gupta, N., Fisker, N., Asselin, M. C., Lindholm, M., Rosenbohm, C., Ørum, H., et al. (2010). A locked nucleic acid antisense oligonucleotide (LNA) silences pcsk9 and enhances ldlr expression in vitro and in vivo. *PLoS One, 5*, e10682.

Guterstam, P., Lindgren, M., Johansson, H., Tedebark, U., Wengel, J., El Andaloussi, S., et al. (2008). Splice-switching efficiency and specificity for oligonucleotides with locked nucleic acid monomers. *The Biochemical Journal, 412*, 307–313.

Hansen, J. B., Fisker, N., Westergaard, M., Kjaerulff, L. S., Hansen, H. F., Thrue, C. A., et al. (2008). Spc3042: A proapoptotic survivin inhibitor. *Molecular Cancer Therapeutics, 7*, 2736–2745.

Heil, F., Hemmi, H., Hochrein, H., Ampenberger, F., Kirschning, C., Akira, S., et al. (2004). Species-specific recognition of single-stranded RNA via toll-like receptor 7 and 8. *Science, 303*, 1526–1529.

Hernandez, F. J., Kalra, N., Wengel, J., & Vester, B. (2009). Aptamers as a model for functional evaluation of LNA and 2'-amino LNA. *Bioorganic & Medicinal Chemistry Letters, 19*, 6585–6587.

Hertoghs, K. M., Ellis, J. H., & Catchpole, I. R. (2003). Use of locked nucleic acid oligonucleotides to add functionality to plasmid DNA. *Nucleic Acids Research, 31*, 5817–5830.

Hildebrandt-Eriksen, E. S., Aarup, V., Persson, R., Hansen, H. F., Munk, M. E., & Ørum, H. (2012). A locked nucleic acid oligonucleotide targeting microRNA 122 is well-tolerated in cynomolgus monkeys. *Nucleic Acid Therapeutics, 22*, 152–161.

Højland, T., Babu, B. R., Bryld, T., & Wengel, J. (2007). Triplex-forming ability of modified oligonucleotides. *Nucleosides, Nucleotides & Nucleic Acids, 26*, 1411–1414.

Højland, T., Kumar, S., Babu, B. R., Umemoto, T., Albaek, N., Sharma, P. K., et al. (2007). LNA (locked nucleic acid) and analogs as triplex-forming oligonucleotides. *Organic & Biomolecular Chemistry, 5*, 2375–2379.

Hong, J., Huang, Y., Li, J., Yi, F., Zheng, J., Huang, H., et al. (2010). Comprehensive analysis of sequence-specific stability of siRNA. *The FASEB Journal, 24*, 4844–4855.

Hornung, V., Guenthner-Biller, M., Bourquin, C., Ablasser, A., Schlee, M., Uematsu, S., et al. (2005). Sequence-specific potent induction of ifn-alpha by short interfering RNA in plasmacytoid dendritic cells through TLR7. *Nature Medicine, 11*, 263–270.

Hrdlicka, P. J., Babu, B. R., Sørensen, M. D., Harrit, N., & Wengel, J. (2005). Multilabeled pyrene-functionalized 2'-amino-LNA probes for nucleic acid detection in homogeneous fluorescence assays. *Journal of the American Chemical Society, 127*, 13293–13299.

Hrdlicka, P. J., Babu, B. R., Sørensen, M. D., & Wengel, J. (2004). Interstrand communication between 2'-n-(pyren-1-yl)methyl-2'-amino-LNA monomers in nucleic acid duplexes: Directional control and signalling of full complementarity. *Chemical Communications (Cambridge, England)*, 1478–1479.

Hummelshoj, L., Ryder, L. P., Madsen, H. O., & Poulsen, L. K. (2005). Locked nucleic acid inhibits amplification of contaminating DNA in real-time pcr. *Biotechniques, 38*, 605–610.

Ittig, D., Liu, S., Renneberg, D., Schumperli, D., & Leumann, C. J. (2004). Nuclear antisense effects in cyclophilin a pre-mRNA splicing by oligonucleotides: A comparison of tricyclo-DNA with LNA. *Nucleic Acids Research, 32*, 346–353.

Janssen, H. L., Reesink, H. W., Lawitz, E. J., Zeuzem, S., Rodriguez-Torres, M., Patel, K., et al. (2013). Treatment of HCV Infection by Targeting MicroRNA. *N Engl J Med.* Mar 27. [Epub ahead of print].

Johannsen, M. W., Crispino, L., Wamberg, M. C., Kalra, N., & Wengel, J. (2011). Amino acids attached to 2'-amino-LNA: Synthesis and excellent duplex stability. *Organic & Biomolecular Chemistry, 9*, 243–252.

Jopling, C. L., Yi, M., Lancaster, A. M., Lemon, S. M., & Sarnow, P. (2005). Modulation of hepatitis c virus RNA abundance by a liver-specific microRNA. *Science, 309*, 1577–1581.

Josefsen, M. H., Lofstrom, C., Sommer, H. M., & Hoorfar, J. (2009). Diagnostic pcr: Comparative sensitivity of four probe chemistries. *Molecular and Cellular Probes, 23,* 201–203.

Judge, A. D., Sood, V., Shaw, J. R., Fang, D., McClintock, K., & MacLachlan, I. (2005). Sequence-dependent stimulation of the mammalian innate immune response by synthetic siRNA. *Nature Biotechnology, 23,* 457–462.

Jurk, M., Chikh, G., Schulte, B., Kritzler, A., Richardt-Pargmann, D., Lampron, C., et al. (2011). Immunostimulatory potential of silencing RNAs can be mediated by a nonuridine-rich toll-like receptor 7 motif. *Nucleic Acid Therapeutics, 21,* 201–214.

Kalish, J. M., Seidman, M. M., Weeks, D. L., & Glazer, P. M. (2005). Triplex-induced recombination and repair in the pyrimidine motif. *Nucleic Acids Research, 33,* 3492–3502.

Kang, S. H., Cho, M. J., & Kole, R. (1998). Up-regulation of luciferase gene expression with antisense oligonucleotides: Implications and applications in functional assay development. *Biochemistry, 37,* 6235–6239.

Kaur, H., Babu, B. R., & Maiti, S. (2007). Perspectives on chemistry and therapeutic applications of locked nucleic acid (LNA). *Chemical Reviews, 107,* 4672–4697.

Kaur, H., Wengel, J., & Maiti, S. (2007). LNA-modified oligonucleotides effectively drive intramolecular-stable hairpin to intermolecular-duplex state. *Biochemical and Biophysical Research Communications, 352,* 118–122.

Khvorova, A., Reynolds, A., & Jayasena, S. D. (2003). Functional siRNAs and miRNAs exhibit strand bias. *Cell, 115,* 209–216.

Kloosterman, W. P., Wienholds, E., de Bruijn, E., Kauppinen, S., & Plasterk, R. H. (2006). In situ detection of miRNAs in animal embryos using LNA-modified oligonucleotide probes. *Nature Methods, 3,* 27–29.

Koch, T., & Ørum, H. (2007). Locked nucleic acid. In S. T. Crooke (Ed.), *Antisense drug discovery* (pp. 519–565). Boca Raton: CRC Press.

Koch, T., Rosenbohm, C., Hansen, B., Straarup, E. M., & Kauppinen, S. (2008). Locked nucleic acid: Properties and therapeutic aspects. In J. Kurreck (Ed.), *Therapeutic oligonucleotide* (pp. 103–141). Cambridge, UK: RSC Publishing.

Koizumi, M., Morita, K., Daigo, M., Tsutsumi, S., Abe, K., Obika, S., et al. (2003). Triplex formation with 2′-o,4′-c-ethylene-bridged nucleic acids (ena) having c3′-endo conformation at physiological ph. *Nucleic Acids Research, 31,* 3267–3273.

Kole, R., Krainer, A. R., & Altman, S. (2012). RNA therapeutics: Beyond RNA interference and antisense oligonucleotides. *Nature Reviews. Drug Discovery, 11,* 125–140.

Koller, E., Vincent, T. M., Chappell, A., De, S., Manoharan, M., & Bennett, C. F. (2011). Mechanisms of single-stranded phosphorothioate modified antisense oligonucleotide accumulation in hepatocytes. *Nucleic Acids Research, 39,* 4795–4807.

Kore, A. R., Hodeib, M., & Hu, Z. (2008). Chemical synthesis of LNA-mctp and its application for microRNA detection. *Nucleosides, Nucleotides & Nucleic Acids, 27,* 1–17.

Koshkin, A. A., Nielsen, P., Meldgaard, M., Rajwanshi, V. K., Singh, S. K., & Wengel, J. (1998). LNA (locked nucleic acid): An RNA mimic forming exceedingly stable LNA: LNA duplexes. *Journal of the American Chemical Society, 120,* 13252–13253.

Krutzfeldt, J., Rajewsky, N., Braich, R., Rajeev, K. G., Tuschl, T., Manoharan, M., et al. (2005). Silencing of microRNAs in vivo with 'antagomirs'. *Nature, 438,* 685–689.

Kumar, N., Nielsen, K. E., Maiti, S., & Petersen, M. (2006). Triplex formation with alpha-l-LNA (alpha-l-ribo-configured locked nucleic acid). *Journal of the American Chemical Society, 128,* 14–15.

Kumar, R., Singh, S. K., Koshkin, A. A., Rajwanshi, V. K., Meldgaard, M., & Wengel, J. (1998). The first analogues of LNA (locked nucleic acids): Phosphorothioate-LNA and 2′-thio-LNA. *Bioorganic & Medicinal Chemistry Letters, 8,* 2219–2222.

Kurreck, J., Wyszko, E., Gillen, C., & Erdmann, V. A. (2002). Design of antisense oligonucleotides stabilized by locked nucleic acids. *Nucleic Acids Research, 30,* 1911–1918.

Kuwahara, M., Obika, S., Nagashima, J., Ohta, Y., Suto, Y., Ozaki, H., et al. (2008). Systematic analysis of enzymatic DNA polymerization using oligo-DNA templates and triphosphate analogs involving $2',4'$-bridged nucleosides. *Nucleic Acids Research, 36,* 4257–4265.

Lanford, R. E., Hildebrandt-Eriksen, E. S., Petri, A., Persson, R., Lindow, M., Munk, M. E., et al. (2010). Therapeutic silencing of microRNA-122 in primates with chronic hepatitis c virus infection. *Science, 327,* 198–201.

Larsen, H. J., Bentin, T., & Nielsen, P. E. (1999). Antisense properties of peptide nucleic acid. *Biochimica et Biophysica Acta, 1489,* 159–166.

Latorra, D., Arar, K., & Hurley, J. M. (2003). Design considerations and effects of LNA in pcr primers. *Molecular and Cellular Probes, 17,* 253–259.

Laufer, S. D., Recke, A. L., Veldhoen, S., Trampe, A., & Restle, T. (2009). Noncovalent peptide-mediated delivery of chemically modified steric block oligonucleotides promotes splice correction: Quantitative analysis of uptake and biological effect. *Oligonucleotides, 19,* 63–80.

Laursen, M. B., Pakula, M. M., Gao, S., Fluiter, K., Mook, O. R., Baas, F., et al. (2010). Utilization of unlocked nucleic acid (una) to enhance siRNA performance in vitro and in vivo. *Molecular BioSystems, 6,* 862–870.

Laxton, C., Brady, K., Moschos, S., Turnpenny, P., Rawal, J., Pryde, D. C., et al. (2011). Selection, optimization, and pharmacokinetic properties of a novel, potent antiviral locked nucleic acid-based antisense oligomer targeting hepatitis c virus internal ribosome entry site. *Antimicrobial Agents and Chemotherapy, 55,* 3105–3114.

Lebedeva, I., Benimetskaya, L., Stein, C. A., & Vilenchik, M. (2000). Cellular delivery of antisense oligonucleotides. *European Journal of Pharmaceutics and Biopharmaceutics, 50,* 101–119.

Levin, J. D., Fiala, D., Samala, M. F., Kahn, J. D., & Peterson, R. J. (2006). Position-dependent effects of locked nucleic acid (LNA) on DNA sequencing and pcr primers. *Nucleic Acids Research, 34,* e142.

Lindholm, M. W., Elmen, J., Fisker, N., Hansen, H. F., Persson, R., Moller, M. R., et al. (2012). Pcsk9 LNA antisense oligonucleotides induce sustained reduction of ldl cholesterol in nonhuman primates. *Molecular Therapy, 20,* 376–381.

Ling, J. Q., Hou, A., & Hoffman, A. R. (2011). Long-range DNA interactions are specifically altered by locked nucleic acid-targeting of a ctcf binding site. *Biochimica et Biophysica Acta, 1809,* 24–33.

Lundin, K. E., Ge, R., Svahn, M. G., Tornquist, E., Leijon, M., Branden, L. J., et al. (2004). Cooperative strand invasion of supercoiled plasmid DNA by mixed linear pna and pna-peptide chimeras. *Biomolecular Engineering, 21,* 51–59.

Lundin, K. E., Good, L., Stromberg, R., Graslund, A., & Smith, C. I. (2006). Biological activity and biotechnological aspects of peptide nucleic acid. *Advances in Genetics, 56,* 1–51.

Machlin, E. S., Sarnow, P., & Sagan, S. M. (2011). Masking the $5'$ terminal nucleotides of the hepatitis c virus genome by an unconventional microRNA-target RNA complex. *Proceedings of the National Academy of Sciences of the United States of America, 108,* 3193–3198.

Mallikaratchy, P. R., Ruggiero, A., Gardner, J. R., Kuryavyi, V., Maguire, W. F., Heaney, M. L., et al. (2011). A multivalent DNA aptamer specific for the b-cell receptor on human lymphoma and leukemia. *Nucleic Acids Research, 39,* 2458–2469.

Manoharan, M. (2002). Oligonucleotide conjugates as potential antisense drugs with improved uptake, biodistribution, targeted delivery, and mechanism of action. *Antisense & Nucleic Acid Drug Development, 12,* 103–128.

Marin, R. M., & Vanicek, J. (2011). Efficient use of accessibility in microRNA target prediction. *Nucleic Acids Research, 39,* 19–29.

Martinez, K., Estevez, M. C., Wu, Y., Phillips, J. A., Medley, C. D., & Tan, W. (2009). Locked nucleic acid based beacons for surface interaction studies and biosensor development. *Analytical Chemistry, 81*, 3448–3454.

Matranga, C., Tomari, Y., Shin, C., Bartel, D. P., & Zamore, P. D. (2005). Passenger-strand cleavage facilitates assembly of siRNA into ago2-containing RNAi enzyme complexes. *Cell, 123*, 607–620.

McTigue, P. M., Peterson, R. J., & Kahn, J. D. (2004). Sequence-dependent thermodynamic parameters for locked nucleic acid (LNA)-DNA duplex formation. *Biochemistry, 43*, 5388–5405.

Moghimi, S. M., Symonds, P., Murray, J. C., Hunter, A. C., Debska, G., & Szewczyk, A. (2005). A two-stage poly(ethylenimine)-mediated cytotoxicity: Implications for gene transfer/therapy. *Molecular Therapy, 11*, 990–995.

Mook, O. R., Baas, F., de Wissel, M. B., & Fluiter, K. (2007). Evaluation of locked nucleic acid-modified small interfering RNA in vitro and in vivo. *Molecular Cancer Therapeutics, 6*, 833–843.

Mook, O., Vreijling, J., Wengel, S. L., Wengel, J., Zhou, C., Chattopadhyaya, J., et al. (2010). In vivo efficacy and off-target effects of locked nucleic acid (LNA) and unlocked nucleic acid (UNA) modified siRNA and small internally segmented interfering RNA (sisiRNA) in mice bearing human tumor xenografts. *Artificial DNA: PNA & XNA, 1*, 36–44.

Moreno, P. M., Geny, S., Pabon, Y. V., Bergquist, H., Zaghloul, E. M., Rocha, C., et al. (2013). Development of bis-locked nucleic acid (bisLNA) oligonucleotides for efficient invasion of supercoiled duplex DNA. *Nucleic Acids Research, 41*, 3257–3273.

Moreno, P. M., Wenska, M., Lundin, K. E., Wrange, O., Stromberg, R., & Smith, C. I. (2009). A synthetic snRNA m3G-cap enhances nuclear delivery of exogenous proteins and nucleic acids. *Nucleic Acids Research, 37*, 1925–1935.

Morita, K., Hasegawa, C., Kaneko, M., Tsutsumi, S., Sone, J., Ishikawa, T., et al. (2002). 2'-o,4'-c-ethylene-bridged nucleic acids (ENA): Highly nuclease-resistant and thermodynamically stable oligonucleotides for antisense drug. *Bioorganic & Medicinal Chemistry Letters, 12*, 73–76.

Morita, K., Takagi, M., Hasegawa, C., Kaneko, M., Tsutsumi, S., Sone, J., et al. (2003). Synthesis and properties of 2'-o,4'-c-ethylene-bridged nucleic acids (ena) as effective antisense oligonucleotides. *Bioorganic & Medicinal Chemistry, 11*, 2211–2226.

Morrissey, D. V., Lockridge, J. A., Shaw, L., Blanchard, K., Jensen, K., Breen, W., et al. (2005). Potent and persistent in vivo anti-hbv activity of chemically modified siRNAs. *Nature Biotechnology, 23*, 1002–1007.

Moschos, S. A., Frick, M., Taylor, B., Turnpenny, P., Graves, H., Spink, K. G., et al. (2011). Uptake, efficacy, and systemic distribution of naked, inhaled short interfering RNA (siRNA) and locked nucleic acid (LNA) antisense. *Molecular Therapy, 19*, 2163–2168.

Nagahama, K., Veedu, R. N., & Wengel, J. (2009). Nuclease resistant methylphosphonate-DNA/LNA chimeric oligonucleotides. *Bioorganic & Medicinal Chemistry Letters, 19*, 2707–2709.

Nielsen, P. E. (2010). Sequence-selective targeting of duplex DNA by peptide nucleic acids. *Current Opinion in Molecular Therapeutics, 12*, 184–191.

Nielsen, P. E., Egholm, M., Berg, R. H., & Buchardt, O. (1991). Sequence-selective recognition of DNA by strand displacement with a thymine-substituted polyamide. *Science, 254*, 1497–1500.

Nielsen, K. E., Rasmussen, J., Kumar, R., Wengel, J., Jacobsen, J. P., & Petersen, M. (2004). Nmr studies of fully modified locked nucleic acid (LNA) hybrids: Solution structure of an LNA:RNA hybrid and characterization of an LNA:DNA hybrid. *Bioconjugate Chemistry, 15*, 449–457.

Nielsen, J. T., Stein, P. C., & Petersen, M. (2003). Nmr structure of an alpha-l-LNA:RNA hybrid: Structural implications for RNAse h recognition. *Nucleic Acids Research, 31,* 5858–5867.

Noir, R., Kotera, M., Pons, B., Remy, J. S., & Behr, J. P. (2008). Oligonucleotide-oligospermine conjugates (zip nucleic acids): A convenient means of finely tuning hybridization temperatures. *Journal of the American Chemical Society, 130,* 13500–13505.

Obad, S., dos Santos, C. O., Petri, A., Heidenblad, M., Broom, O., Ruse, C., et al. (2011). Silencing of microRNA families by seed-targeting tiny LNAs. *Nature Genetics, 43,* 371–378.

Obika, S. (2004). Development of bridged nucleic acid analogues for antigene technology. *Chemical & Pharmaceutical Bulletin (Tokyo), 52,* 1399–1404.

Obika, S., Nanbu, D., Hari, Y., Andoh, J.-i., Morio, K.-i., Doi, T., et al. (1998). Stability and structural features of the duplexes containing nucleoside analogues with a fixed n-type conformation, 2′-o,4′- c-methyleneribonucleosides. *Tetrahedron Letters, 39,* 5401–5404.

Obika, S., Nanbu, D., Hari, Y., Morio, K.-I., In, Y., Ishida, T., et al. (1997). Synthesis of 2′-o,4′-c-methyleneuridine and -cytidine. Novel bicyclic nucleosides having a fixed c3′-endo sugar puckering. *Tetrahedron Letters, 38,* 8735–8738.

Obika, S., Rahman, S. M. A., Fujisaka, A., Kawada, Y., Baba, T., & Imanishi, T. (2010). Bridged nucleic acids: Development, synthesis and properties. *Heterocycles, 81,* 1347–1392.

Obika, S., Uneda, T., Sugimoto, T., Nanbu, D., Minami, T., Doi, T., et al. (2001). 2′-o, 4′-c-methylene bridged nucleic acid (2′,4′-bna): Synthesis and triplex-forming properties. *Bioorganic & Medicinal Chemistry, 9,* 1001–1011.

Overhoff, M., & Sczakiel, G. (2005). Phosphorothioate-stimulated uptake of short interfering RNA by human cells. *EMBO Reports, 6,* 1176–1181.

Park, E., Gang, E. J., Hsieh, Y. T., Schaefer, P., Chae, S., Klemm, L., et al. (2011). Targeting survivin overcomes drug resistance in acute lymphoblastic leukemia. *Blood, 118,* 2191–2199.

Pasternak, K., Pasternak, A., Gupta, P., Veedu, R. N., & Wengel, J. (2011). Photoligation of self-assembled DNA constructs containing anthracene-functionalized 2′-amino-LNA monomers. *Bioorganic & Medicinal Chemistry, 19,* 7407–7415.

Petersen, M., Bondensgaard, K., Wengel, J., & Jacobsen, J. P. (2002). Locked nucleic acid (LNA) recognition of RNA: Nmr solution structures of LNA:RNA hybrids. *Journal of the American Chemical Society, 124,* 5974–5982.

Petersen, M., Nielsen, C. B., Nielsen, K. E., Jensen, G. A., Bondensgaard, K., Singh, S. K., et al. (2000). The conformations of locked nucleic acids (LNA). *Journal of Molecular Recognition, 13,* 44–53.

Pinheiro, V. B., Taylor, A. I., Cozens, C., Abramov, M., Renders, M., Zhang, S., et al. (2012). Synthetic genetic polymers capable of heredity and evolution. *Science, 336,* 341–344.

Proske, D., Blank, M., Buhmann, R., & Resch, A. (2005). Aptamers—Basic research, drug development, and clinical applications. *Applied Microbiology and Biotechnology, 69,* 367–374.

Rahman, S. M., Seki, S., Obika, S., Yoshikawa, H., Miyashita, K., & Imanishi, T. (2008). Design, synthesis, and properties of 2′,4′-bna(nc): A bridged nucleic acid analogue. *Journal of the American Chemical Society, 130,* 4886–4896.

Rand, T. A., Petersen, S., Du, F., & Wang, X. (2005). Argonaute2 cleaves the anti-guide strand of siRNA during risc activation. *Cell, 123,* 621–629.

Renneberg, D., Bouliong, E., Reber, U., Schumperli, D., & Leumann, C. J. (2002). Antisense properties of tricyclo-DNA. *Nucleic Acids Research, 30,* 2751–2757.

Renneberg, D., & Leumann, C. J. (2002). Watson-crick base-pairing properties of tricyclo-DNA. *Journal of the American Chemical Society, 124,* 5993–6002.

Reyes-Reyes, E. M., Teng, Y., & Bates, P. J. (2010). A new paradigm for aptamer therapeutic as1411 action: Uptake by macropinocytosis and its stimulation by a nucleolin-dependent mechanism. *Cancer Research*, *70*, 8617–8629.

Reynisson, E., Josefsen, M. H., Krause, M., & Hoorfar, J. (2006). Evaluation of probe chemistries and platforms to improve the detection limit of real-time pcr. *Journal of Microbiological Methods*, *66*, 206–216.

Reynolds, A., Anderson, E. M., Vermeulen, A., Fedorov, Y., Robinson, K., Leake, D., et al. (2006). Induction of the interferon response by siRNA is cell type- and duplex length-dependent. *RNA*, *12*, 988–993.

Roberts, J., Palma, E., Sazani, P., Ørum, H., Cho, M., & Kole, R. (2006). Efficient and persistent splice switching by systemically delivered LNA oligonucleotides in mice. *Molecular Therapy*, *14*, 471–475.

Rockwell, P., O'Connor, W. J., King, K., Goldstein, N. I., Zhang, L. M., & Stein, C. A. (1997). Cell-surface perturbations of the epidermal growth factor and vascular endothelial growth factor receptors by phosphorothioate oligodeoxynucleotides. *Proceedings of the National Academy of Sciences of the United States of America*, *94*, 6523–6528.

Sano, M., Sierant, M., Miyagishi, M., Nakanishi, M., Takagi, Y., & Sutou, S. (2008). Effect of asymmetric terminal structures of short RNA duplexes on the RNA interference activity and strand selection. *Nucleic Acids Research*, *36*, 5812–5821.

Sapra, P., Wang, M., Bandaru, R., Zhao, H., Greenberger, L. M., & Horak, I. D. (2010). Down-modulation of survivin expression and inhibition of tumor growth in vivo by ezn-3042, a locked nucleic acid antisense oligonucleotide. *Nucleosides, Nucleotides & Nucleic Acids*, *29*, 97–112.

Sasaki, K., Rahman, S. M., Sato, N., Obika, S., Imanishi, T., & Torigoe, H. (2009). Promotion of triplex formation by 3′-amino-2′-o,4′-c-methylene bridged nucleic acid modification. *Nucleic Acids Symposium Series*, *53*, 159–160.

Sau, S. P., Kumar, P., Anderson, B. A., Ostergaard, M. E., Deobald, L., Paszczynski, A., et al. (2009). Optimized DNA-targeting using triplex forming c5-alkynyl functionalized LNA. *Chemical Communications (Cambridge, England)*, *44*, 6756–6758.

Sau, S. P., Kumar, T. S., & Hrdlicka, P. J. (2010). Invader LNA: Efficient targeting of short double stranded DNA. *Organic & Biomolecular Chemistry*, *8*, 2028–2036.

Schmidt, K. S., Borkowski, S., Kurreck, J., Stephens, A. W., Bald, R., Hecht, M., et al. (2004). Application of locked nucleic acids to improve aptamer in vivo stability and targeting function. *Nucleic Acids Research*, *32*, 5757–5765.

Schwarz, D. S., Hutvagner, G., Du, T., Xu, Z., Aronin, N., & Zamore, P. D. (2003). Asymmetry in the assembly of the RNAi enzyme complex. *Cell*, *115*, 199–208.

Schyth, B. D., Bramsen, J. B., Pakula, M. M., Larashati, S., Kjems, J., Wengel, J., et al. (2012). In vivo screening of modified siRNAs for non-specific antiviral effect in a small fish model: Number and localization in the strands are important. *Nucleic Acids Research*, *40*, 4653–4665.

Shabalina, S. A., Spiridonov, A. N., & Ogurtsov, A. Y. (2006). Computational models with thermodynamic and composition features improve siRNA design. *BMC Bioinformatics*, *7*, 65.

Shangguan, D., Tang, Z., Mallikaratchy, P., Xiao, Z., & Tan, W. (2007). Optimization and modifications of aptamers selected from live cancer cell lines. *ChemBioChem*, *8*, 603–606.

Shen, L., Johnson, T. L., Clugston, S., Huang, H., Butenhof, K. J., & Stanton, R. V. (2011). Molecular dynamics simulation and binding energy calculation for estimation of oligonucleotide duplex thermostability in RNA-based therapeutics. *Journal of Chemical Information and Modeling*, *51*, 1957–1965.

Shi, F., Gounko, N. V., Wang, X., Ronken, E., & Hoekstra, D. (2007). In situ entry of oligonucleotides into brain cells can occur through a nucleic acid channel. *Oligonucleotides*, *17*, 122–133.

Shibahara, S., Mukai, S., Morisawa, H., Nakashima, H., Kobayashi, S., & Yamamoto, N. (1989). Inhibition of human immunodeficiency virus (hiv-1) replication by synthetic oligo-RNA derivatives. *Nucleic Acids Research, 17*, 239–252.

Shimakami, T., Yamane, D., Jangra, R. K., Kempf, B. J., Spaniel, C., Barton, D. J., et al. (2012). Stabilization of hepatitis c virus RNA by an ago2-mir-122 complex. *Proceedings of the National Academy of Sciences of the United States of America, 109*, 941–946.

Silahtaroglu, A., Pfundheller, H., Koshkin, A., Tommerup, N., & Kauppinen, S. (2004). LNA-modifed oligonucleotides are highly efficient as fish probes. *Cytogenetic and Genome Research, 107*, 32–37.

Simoes-Wust, A. P., Hopkins-Donaldson, S., Sigrist, B., Belyanskaya, L., Stahel, R. A., & Zangemeister-Wittke, U. (2004). A functionally improved locked nucleic acid antisense oligonucleotide inhibits bcl-2 and bcl-xl expression and facilitates tumor cell apoptosis. *Oligonucleotides, 14*, 199–209.

Singh, S. K., Kumar, R., & Wengel, J. (1998a). Synthesis of 2′-amino-LNA: A novel conformationally restricted high-affinity oligonucleotide analogue with a handle. *The Journal of Organic Chemistry, 63*, 10035–10039.

Singh, S. K., Kumar, R., & Wengel, J. (1998b). Synthesis of novel bicyclo[2.2.1] ribonucleosides: 2′-amino- and 2′-thio-LNA monomeric nucleosides. *The Journal of Organic Chemistry, 63*, 6078–6079.

Singh, S. K., Nielsen, P., Koshkin, A. A., & Wengel, J. (1998). LNA (locked nucleic acids): Synthesis and high-affinity nucleic acid recognition. *Chemical Communications (Cambridge, England), 4*, 455–456.

Sioud, M. (2005). Induction of inflammatory cytokines and interferon responses by double-stranded and single-stranded siRNAs is sequence-dependent and requires endosomal localization. *Journal of Molecular Biology, 348*, 1079–1090.

Sioud, M. (2006). Single-stranded small interfering RNA are more immunostimulatory than their double-stranded counterparts: A central role for 2′-hydroxyl uridines in immune responses. *European Journal of Immunology, 36*, 1222–1230.

Sledz, C. A., Holko, M., de Veer, M. J., Silverman, R. H., & Williams, B. R. (2003). Activation of the interferon system by short-interfering RNAs. *Nature Cell Biology, 5*, 834–839.

Sørensen, M. D., Kvaerno, L., Bryld, T., Hakansson, A. E., Verbeure, B., Gaubert, G., et al. (2002). Alpha-l-ribo-configured locked nucleic acid (alpha-l-LNA): Synthesis and properties. *Journal of the American Chemical Society, 124*, 2164–2176.

Sorrentino, S. (1998). Human extracellular ribonucleases: Multiplicity, molecular diversity and catalytic properties of the major RNAse types. *Cellular and Molecular Life Sciences, 54*, 785–794.

Soutschek, J., Akinc, A., Bramlage, B., Charisse, K., Constien, R., Donoghue, M., et al. (2004). Therapeutic silencing of an endogenous gene by systemic administration of modified siRNAs. *Nature, 432*, 173–178.

Spitali, P., & Aartsma-Rus, A. (2012). Splice modulating therapies for human disease. *Cell, 148*, 1085–1088.

Stein, C. A., Hansen, J. B., Lai, J., Wu, S., Voskresenskiy, A., Hog, A., et al. (2010). Efficient gene silencing by delivery of locked nucleic acid antisense oligonucleotides, unassisted by transfection reagents. *Nucleic Acids Research, 38*, e3.

Stenvang, J., Petri, A., Lindow, M., Obad, S., & Kauppinen, S. (2012). Inhibition of microRNA function by antimir oligonucleotides. *Silence, 3*, 1.

Stenvang, J., Silahtaroglu, A. N., Lindow, M., Elmen, J., & Kauppinen, S. (2008). The utility of LNA in microRNA-based cancer diagnostics and therapeutics. *Seminars in Cancer Biology, 18*, 89–102.

Stirchak, E. P., Summerton, J. E., & Weller, D. D. (1989). Uncharged stereoregular nucleic acid analogs: 2. Morpholino nucleoside oligomers with carbamate internucleoside linkages. *Nucleic Acids Research, 17*, 6129–6141.

Straarup, E. M., Fisker, N., Hedtjarn, M., Lindholm, M. W., Rosenbohm, C., Aarup, V., et al. (2010). Short locked nucleic acid antisense oligonucleotides potently reduce apolipoprotein b mRNA and serum cholesterol in mice and non-human primates. *Nucleic Acids Research*, *38*, 7100–7111.

Sun, B. W., Babu, B. R., Sørensen, M. D., Zakrzewska, K., Wengel, J., & Sun, J. S. (2004). Sequence and ph effects of LNA-containing triple helix-forming oligonucleotides: Physical chemistry, biochemistry, and modeling studies. *Biochemistry*, *43*, 4160–4169.

Sun, Z., Xiang, W., Guo, Y., Chen, Z., Liu, W., & Lu, D. (2011). Inhibition of hepatitis b virus (hbv) by LNA-mediated nuclear interference with hbv DNA transcription. *Biochemical and Biophysical Research Communications*, *409*, 430–435.

Syed, M. A., & Pervaiz, S. (2010). Advances in aptamers. *Oligonucleotides*, *20*, 215–224.

Symonds, P., Murray, J. C., Hunter, A. C., Debska, G., Szewczyk, A., & Moghimi, S. M. (2005). Low and high molecular weight poly(l-lysine)s/poly(l-lysine)-DNA complexes initiate mitochondrial-mediated apoptosis differently. *FEBS Letters*, *579*, 6191–6198.

Takahashi, M., Nagai, C., Hatakeyama, H., Minakawa, N., Harashima, H., & Matsuda, A. (2012). Intracellular stability of 2′-OMe-4′-thioribonucleoside modified siRNA leads to long-term RNAi effect. *Nucleic Acids Research*, *40*, 5787–5793.

Thorsen, S. B., Obad, S., Jensen, N. F., Stenvang, J., & Kauppinen, S. (2012). The therapeutic potential of microRNAs in cancer. *Cancer Journal*, *18*, 275–284.

Torigoe, H., Hari, Y., Sekiguchi, M., Obika, S., & Imanishi, T. (2001). 2′-o,4′-c-methylene bridged nucleic acid modification promotes pyrimidine motif triplex DNA formation at physiological ph: Thermodynamic and kinetic studies. *The Journal of Biological Chemistry*, *276*, 2354–2360.

Torigoe, H., Sato, N., & Nagasawa, N. (2012). 2′-o,4′-c-ethylene bridged nucleic acid modification enhances pyrimidine motif triplex-forming ability under physiological condition. *Journal of Biochemistry*, *152*, 17–26.

Ugozzoli, L. A., Latorra, D., Puckett, R., Arar, K., & Hamby, K. (2004). Real-time genotyping with oligonucleotide probes containing locked nucleic acids. *Analytical Biochemistry*, *324*, 143–152.

Umemoto, T., Wengel, J., & Madsen, A. S. (2009). Functionalization of 2′-amino-LNA with additional nucleobases. *Organic & Biomolecular Chemistry*, *7*, 1793–1797.

Veedu, R. N., Vester, B., & Wengel, J. (2007a). Enzymatic incorporation of LNA nucleotides into DNA strands. *ChemBioChem*, *8*, 490–492.

Veedu, R. N., Vester, B., & Wengel, J. (2007b). In vitro incorporation of LNA nucleotides. *Nucleosides, Nucleotides & Nucleic Acids*, *26*, 1207–1210.

Veedu, R. N., Vester, B., & Wengel, J. (2008). Polymerase chain reaction and transcription using locked nucleic acid nucleotide triphosphates. *Journal of the American Chemical Society*, *130*, 8124–8125.

Veedu, R. N., Vester, B., & Wengel, J. (2009). Efficient enzymatic synthesis of LNA-modified DNA duplexes using kod DNA polymerase. *Organic & Biomolecular Chemistry*, *7*, 1404–1409.

Veedu, R. N., Vester, B., & Wengel, J. (2010). Polymerase directed incorporation studies of LNA-G nucleoside 5′-triphosphate and primer extension involving all four LNA nucleotides. *New Journal of Chemistry*, *34*, 877–879.

Veedu, R. N., & Wengel, J. (2010). Locked nucleic acids: Promising nucleic acid analogs for therapeutic applications. *Chemistry & Biodiversity*, *7*, 536–542.

Virno, A., Randazzo, A., Giancola, C., Bucci, M., Cirino, G., & Mayol, L. (2007). A novel thrombin binding aptamer containing a g-LNA residue. *Bioorganic & Medicinal Chemistry*, *15*, 5710–5718.

Volkov, A. A., Kruglova, N. S., Meschaninova, M. I., Venyaminova, A. G., Zenkova, M. A., Vlassov, V. V., et al. (2009). Selective protection of nuclease-sensitive sites in siRNA prolongs silencing effect. *Oligonucleotides*, *19*, 191–202.

Wahl, M. C., Will, C. L., & Luhrmann, R. (2009). The spliceosome: Design principles of a dynamic rnp machine. *Cell*, *136*, 701–718.

Wahlestedt, C., Salmi, P., Good, L., Kela, J., Johnsson, T., Hokfelt, T., et al. (2000). Potent and nontoxic antisense oligonucleotides containing locked nucleic acids. *Proceedings of the National Academy of Sciences of the United States of America*, *97*, 5633–5638.

Wang, L., Yang, C. J., Medley, C. D., Benner, S. A., & Tan, W. (2005). Locked nucleic acid molecular beacons. *Journal of the American Chemical Society*, *127*, 15664–15665.

Wengel, J., Koshkin, A., Singh, S. K., Nielsen, P., Meldgaard, M., Rajwanshi, V. K., et al. (1999). LNA (locked nucleic acid). *Nucleosides, Nucleotides & Nucleic Acids*, *18*, 1365–1370.

Werk, D., Wengel, J., Wengel, S. L., Grunert, H. P., Zeichhardt, H., & Kurreck, J. (2010). Application of small interfering RNAs modified by unlocked nucleic acid (una) to inhibit the heart-pathogenic coxsackievirus b3. *FEBS Letters*, *584*, 591–598.

Worm, J., Stenvang, J., Petri, A., Frederiksen, K. S., Obad, S., Elmen, J., et al. (2009). Silencing of microRNA-155 in mice during acute inflammatory response leads to derepression of c/ebp beta and down-regulation of g-csf. *Nucleic Acids Research*, *37*, 5784–5792.

Yakovchuk, P., Protozanova, E., & Frank-Kamenetskii, M. D. (2006). Base-stacking and base-pairing contributions into thermal stability of the DNA double helix. *Nucleic Acids Research*, *34*, 564–574.

Yamamoto, T., Nakatani, M., Narukawa, K., & Obika, S. (2011). Antisense drug discovery and development. *Future Medicinal Chemistry*, *3*, 339–365.

Zaghloul, E. M., Madsen, A. S., Moreno, P. M., Oprea, I. I., El-Andaloussi, S., Bestas, B., et al. (2011). Optimizing anti-gene oligonucleotide 'zorro-LNA' for improved strand invasion into duplex DNA. *Nucleic Acids Research*, *39*, 1142–1154.

Zaghloul, E. M., Moreno, P. M., Mohamed, A. J., Wengel, J., Lundin, K. E., & Smith, C. I. (2012). The anti-gene oligonucleotide zorro-LNA, delivered by a cationic lipid, down-regulates huntingtin gene expression in mammalian cells. *Molecular Therapy*, *20*(Suppl. 1), S133–S134.

Zhang, Y., Castaneda, S., Dumble, M., Wang, M., Mileski, M., Qu, Z., et al. (2011). Reduced expression of the androgen receptor by third generation of antisense shows antitumor activity in models of prostate cancer. *Molecular Cancer Therapeutics*, *10*, 2309–2319.

Zhou, C., & Chattopadhyaya, J. (2009). The synthesis of therapeutic locked nucleos(t)ides. *Current Opinion in Drug Discovery & Development*, *12*, 876–898.

INDEX

Note: Page numbers followed by "*f*" indicate figures, and "*t*" indicate tables.